PRENTICE-HALL

FOUNDATIONS OF DEVELOPMENTAL BIOLOGY SERIES

Clement L. Markert, Editor

Volumes published or in preparation:

FERTILIZATION   *C. R. Austin*

CONTROL MECHANISMS IN PLANT DEVELOPMENT
Arthur W. Galston and Peter J. Davies

PRINCIPLES OF MAMMALIAN AGING
Robert R. Kohn

EMBRYONIC DIFFERENTIATION
H. E. Lehman

DEVELOPMENTAL GENETICS*   *Clement L. Markert and Heinrich Ursprung*

CELL REPRODUCTION DURING DEVELOPMENT
David M. Prescott

PATTERNS IN PLANT DEVELOPMENT
T. A. Steeves and I. M. Sussex

CELLS INTO ORGANS: The Forces That Shape
the Embryo   *J. P. Trinkaus*

* Published jointly in Prentice-Hall's *Foundations of Modern Genetics Series*

PATTERNS
IN
PLANT
DEVELOPMENT

Taylor A. Steeves

University of Saskatchewan

Ian M. Sussex

Yale University

PRENTICE-HALL, INC.   Englewood Cliffs, New Jersey

FOUNDATIONS OF DEVELOPMENTAL BIOLOGY SERIES

Printed in the United States of America
ISBN—0-13-653998-X
Library of Congress Catalog Card Number: 77-180599

10   9   8   7   6   5   4   3   2

PRENTICE-HALL INTERNATIONAL, INC., London
PRENTICE-HALL OF AUSTRALIA, PTY. LTD., Sydney
PRENTICE-HALL OF CANADA, LTD., Toronto
PRENTICE-HALL OF INDIA PRIVATE LIMITED, New Delhi
PRENTICE-HALL OF JAPAN, INC., Tokyo

to

C. W. WARDLAW and R. H. WETMORE
whose pioneering research and dedicated teaching have
inspired a generation of developmental botanists

Contents

Foundations of DEVELOPMENTAL BIOLOGY

The development of organisms is so wondrous and yet so common that it has compelled man's attention and aroused his curiosity from earliest times. But developmental processes have proved to be complex and difficult to understand, and little progress was made for hundreds of years. By the beginning of this century, increasingly skillful experimentation began to accelerate the slow advance in our understanding of development. Most important in recent years has been the rapid progress in the related disciplines of biochemistry and genetics—progress that has now made possible an experimental attack on developmental problems at the molecular level. Many old and intractable problems are taking on a fresh appeal, and a tense expectancy pervades the biological community. Rapid advances are surely imminent.

New insights into the structure and function of cells are moving the principal problems of developmental biology into the center of scientific attention, and increasing numbers of biologists are focusing their research efforts on these problems. Moreover, new tools and experimental designs are now available to help in their solution.

At this critical stage of scientific development a fresh assessment is needed. This series of books in developmental biology is designed to provide essential background material and then to examine the frontier where significant advances are occurring or expected. Each book is written by a leading investigator actively concerned with the problems and concepts he discusses. Students at intermediate and advanced levels of preparation and investigators in other areas of biology should find these books informative, stimulating, and useful. Collectively, they present an authoritative and penetrating analysis of the major problems and concepts of developmental biology, together with a critical appraisal of the experimental tools and designs that make developmental biology so exciting and challenging today.

CLEMENT L. MARKERT

Preface

"Hallo!" said Piglet, "what are *you* doing?"
"Hunting," said Pooh.
"Hunting what?"
"Tracking something," said Winnie-the-Pooh very mysteriously.
"Tracking what?" said Piglet, coming closer.
"That's just what I ask myself. I ask myself, What?".
"What do you think you'll answer?"
"I shall have to wait until I catch up with it," said Winnie-the-Pooh.

Winnie-The-Pooh, A. A. Milne*

A skilled hunter usually knows what game he is after, and thus Pooh's bumbling efforts bring a patronizing smile to the lips of the adult reader. The child, however, endowed with that mysterious insight which he most certainly has not acquired from his elders, understands only too well that if you know at the beginning what you are going to find at the end, the fun is over before you start. Scientific enquiry is something like that too, and most researchers would long ago have abandoned their quest in favor of something more challenging if their role were simply to follow well-marked paths to readily forseeable goals. Thus we make no apology for the fact that this book, in which we attempt to outline the present status of the search for an understanding of plant development, does not begin with a detailed plan of where this field is going, does not end with an estimate of our proximity to the ultimate goal, and does not along the way evaluate each major discovery in terms of its contribution to progress towards that goal. Such a plan could be produced from the boundless resources of our preconceived notions; but we recognize only too well that some investigation in progress as these words are being written and yet unknown to us could render such a plan valueless even to us. Our only hope is that we have been able to avoid the mistake committed by Milne's fictional friends, and that we have not simply followed our own footprints.

*From the book *Winnie-the-Pooh* by A. A. Milne. Decorations by E. H. Shepard. Copyright © 1926 by E. P. Dutton & Co., Inc. Renewal 1954 by A. A. Milne. Published by E. P. Dutton & Co., Inc., and used with their permission. Published in Great Britain by Methuen & Co. Ltd., and in Canada by McClelland & Stewart.

The point of view of this volume is, as its title implies, structural. There are two reasons for this. In the first place this is the point of view of the authors in their own teaching and research. The investigation of plant development began as a morphological study; but rapid advances in physiological, biochemical, cytological, and genetic areas have opened broad new approaches to the understanding of developmental phenomena. It is our conviction, however, that these additional approaches in no way diminish the importance of structure. Indeed they make a real grasp of organismal structure even more imperative if control mechanisms are to be understood as biological processes. Secondly, this series on Developmental Biology includes a volume devoted specifically to control mechanisms in plant development. Since both volumes were intended to be used together, it seemed pointless to permit extensive overlap between them.

On the other hand, we hasten to add that we have not attempted to produce merely a descriptive account of structural changes in development. In addition to being of very little interest to us personally, such a treatment would grossly misrepresent the field of developmental morphology as it is today. Increasingly this field of investigation is becoming "causal" in its interpretations; and experimentation has joined description as its tool. Thus the boundary between morphology and physiology becomes less and less clear. We have not analyzed control mechanisms as processes but we have been very much concerned with them as they regulate developmental patterns.

In any attempt to treat an extensive subject briefly, the problem of what to include and what to omit looms large, and the writing of this book has been no exception. Believing as we do that a meaningful understanding of plant development demands an organismal approach, we have elected not to follow the more economical (at least in terms of space) technique of analysing phenomena such as cell growth, meristematic activity, or polarity as topics in themselves. Rather we have attempted, as far as possible, to document the developmental process as the plant undergoes it beginning with the zygote and the formation of the embryo, continuing with the development of the primary body, and completing the picture with a treatment of secondary growth. Certain topics, notably differentiation and the potency of differentiated cells have indeed been given special treatment; but the criticism may be levelled validly that this overall approach does not draw out fundamental generalizations about development which the alternative treatment might have revealed. Nevertheless, we set out to show how the plant develops as an organism and we have endeavored to achieve this goal.

Limitation of space has imposed two restrictions upon this account—one unfortunate, the other perhaps advantageous. Regrettably, we have found it necessary to limit our treatment to the higher or vascular plants and thus have excluded exciting work dealing with algae, fungi, and bryophytes. Even within this limitation, we have further diminished the quantity of descriptive material by selecting particular examples which have been well documented and illustrated in the literature and using these to present particular developmental patterns. The variability in the pattern has then been indicated very briefly following the detailed account of the "type." We believe, however, that this second limitation is a desirable one because it has enabled us to present sufficient detail to provide a meaningful picture without, we hope, too great a burden of description.

Unless we have missed our mark widely, this book ought to be intelligible to any student or other potential user who has completed a first introduction to biology in a university level course. This is an essential feature, because the book is intended to be an introduction to plant development. While avoiding the danger of preparing a text on plant anatomy, we have endeavored to provide enough fundamental information upon which to base an account of developmental phenomena. At the same time, in expanding each topic, we have tried to put the reader quickly in contact with current problems and interpretations. If we have achieved our goal, this introduction to plant development should prepare the reader to continue his studies in the current literature of the field.

Finally we cannot fail to acknowledge our indebtedness to the many who have contributed to this volume by their influence upon us. The two distinguished botanists to whom this book is dedicated were our teachers as we began our own investigations of plant development; and they have continued to guide us through the years by their wise advice and warm encouragement. If there is anything of value here, it can largely be attributed to them. For our errors of fact and judgment, however, they can no longer be held accountable. Many colleagues and students have also contributed to this work, often unwittingly, in numerous discussions and debates on the problems of development. We are especially grateful to the reviewers, in principle unknown to us, to whom the publisher submitted the manuscript for appraisal. Their comments and criticisms have been of great value in correcting our errors and in smoothing out the rough places in our prose. Lastly, we are deeply indebted to the editors of Prentice Hall, Inc., who, while always ready to provide aid and encouragement when it was needed, had the consideration, and perhaps the good sense, to leave us alone to write this book as we saw fit. We hope that their confidence has been merited.

TAYLOR A. STEEVES
IAN M. SUSSEX

ONE

Development in
the Vascular Plants

It is incumbent upon the authors of a book about development to state as precisely as possible what they mean by this term. Development is the sum total of events that contribute to the progressive elaboration of the body of an organism. A more restrictive definition might unfortunately eliminate from consideration some essential aspects of the developmental process. On the other hand, a broad interpretation such as we have given might seem to include all or most of the physiological processes of living organisms. This is especially true in plants in which, in a very real sense, all physiological phenomena throughout the life of the organism seem to be channeled into the progressive elaboration of its body. One might justifiably ask, then, whether such phenomena as photosynthesis or the absorption of water are to be regarded as part of development. Certainly their absence would preclude development; but they are far removed from the actual mechanisms by which the plant body is elaborated.

In order to avoid the impossible requirement of discussing the entire biology of plants, it will be necessary to concentrate upon those phenomena that directly participate in the formation of the plant body. Even in this limited sense, development encompasses numerous processes such as cell division, cell enlargement, protein synthesis, the elaboration of cell wall materials, quantitative and qualitative alterations in cell organelles, and many others. It is, however, convenient to recognize two major aspects of development and to analyze developmental processes in terms of these two categories. These are *growth* and *differentiation*. In an organism, *growth* is an irreversible increase in size, and it is accomplished by a combination of cell division and cell enlargement. Cell division does not of itself constitute growth and in fact may occur without any increase in the over-all size of the structure involved. Cell enlargement alone does constitute growth; this is particularly evident in plants in which there is a considerable net increase in cell size in maturing regions. Nonetheless, with few exceptions the continued growth of an organism requires the production of new cells and their enlargement, and these two processes are closely associated in space and time. Growth by itself will not lead to the formation of an organized body, but rather, at least in theory, to a homogeneous assemblage of cells. Clearly, the formation of an organized body implies that cells and groups of cells in different regions of the body have become structurally distinguishable and functionally distinctive. The changes that occur in these cells and groups of cells and bring about their distinctiveness constitute what is known as *differentiation*. Some biologists prefer to distinguish between those changes that lead to distinctive histological patterns—designated as cell differentiation or histodifferentiation—and those that set apart major segments of the body or organs—designated as organogenesis. Because there is no reason to suppose that mechanisms underlying these two types of changes are fundamentally different, it seems preferable to consider both as aspects of a general phenomenon of differentiation. There are cases in which growth can occur without differentiation and differentiation without growth; but it is almost always true that these two phenomena occur in intimate association. The development of an organized body depends upon the integrated activity of the two.

The vascular plant, like all sexually reproducing organisms, begins its existence as a single cell, the fertilized egg or zygote. Proliferation of this cell leads to the formation of an embryo within which, at an early stage, organs and tissues begin to differentiate. Such differentiation in the animal embryo lays down the fundamental plan of the body; and postembryonic development consists of the enlargement of this body and its maintenance in a functionally efficient state. In the plant embryo, by contrast, the full body plan is not elaborated. Rather, early in embryonic differentiation, two distinctive regions are set apart, approximately at opposite poles, that subsequently retain the capacity for essentially unlimited growth. One of these, designated the *shoot apical meristem*, functions throughout the life of the plant to produce an expanding shoot by the continued formation of tissues and the initiation of a succession of leaf and bud primordia. The other, the *root apical meristem*, similarly forms a potentially unlimited root system. The potentially unlimited growth of the plant body through the activity of apical meristems differentiated in the embryo has been termed *permanent embryogeny*. It is evident also that the development of this open-ended system is repetitive; the same kinds of tissues and organs are produced in continuing succession. Thus, whereas higher animals have characteristic numbers of organs, the indeterminate shoot has an indefinite number of leaves.

The activity of the apical meristems results in the production of a continuously elongating body, which has been called the *primary body* of the plant. In many cases this primary body constitutes the whole plant. In other cases, particularly in those plants with an extended life-span, there is an additional component of development that leads to an increase in girth of the axis. This results from the activity of two additional meristems—the *vascular cambium*, which contributes additional cells to the conducting system, and the *cork cambium*, which produces a protective tissue replacing the original epidermis. These meristems and the tissues they produce constitute what is designated the *secondary body* of the plant. The secondary body does not constitute an entire plant in that it is composed of a few tissue types only and includes no organs. In some cases, however, it comes to constitute the bulk of the plant body.

The possible significance of this mode of development may be appreciated by considering the cellular structure of plants and their over-all immobility. Plant cells are surrounded by a relatively rigid wall and are tightly cemented together within the framework of the tissues. In the animal body, cell replacement, which seems to be essential for the maintenance of funtional efficiency of at least certain tissues, can be accomplished within the histological framework. In the plant, this obviously is impossible, and a comparable replacement is effected by the continual addition of new cells at the growing tips of shoots and roots by the primary meristems and laterally by the secondary meristems. In a very real sense, plants assimilate at their most recently formed tips. The absorption of water and mineral salts is accomplished at or near growing root tips, and shoot apical meristems continually replenish the complement of photosynthetic leaves, which are shed often after one growing season and at most after several years. These expanding tips are connected by a vascular conducting system that may be renewed by cells derived from the vascular cambium. Moreover, continued growth endows the immobile plant with a measure of responsiveness to its environment, which is seen in the tropistic movements of the plant organs and in the continued advance of roots into unoccupied regions of the soil.

There have been many attempts to formulate a comprehensive theoretical basis for the understanding of development. Some of the proposed theories are highly abstract, involving mathematical models or laws of physics. Others are more closely connected to specific developmental phenomena and deal with physiological fields and gradients of metabolites and hormones. Still others combine mathematical and chemical approaches; and the genetic basis of ontogenetic events has been stressed by a number of students of development. In considering the diversity of these theories and the difficulty of applying them on a broad basis, the impression grows that a generalized theory of this complex phenomenon may be premature in the present state of knowledge of development. In this book a rather pragmatic approach will be adopted, namely examination of the facts of development as they are known in the expectation that some concepts of wider application will emerge.

Although a book such as this one ought to contain enough information and interpretation to be useful by itself, it is desirable that readers, particularly students, have access to the original studies in the field. For this reason, at the end of each chapter a selected bibliography is given, and reference is made to it in the context of the chapter. Although these lists are not comprehensive, they contain key reference works that offer the opportunity for a wider grasp of plant development. There are many extremely useful general reference works, books, and major review articles that are basic sources of additional information and points of view. A selected list of these references follows.

REFERENCES

Allsopp, A. 1964. Shoot morphogenesis. Ann. Rev. Plant Physiol. **15**:225–254.

Clowes, F. A. L. 1961. Apical meristems. Blackwell, Oxford.

Clowes, F. A. L., and B. E. Juniper. 1968. Plant cells. Blackwell, Oxford.

Cutter, E. G. 1965. Recent experimental studies of the shoot apex and shoot morphogenesis. Botan. Rev. **31**:7–113.

Esau, K. 1960. Anatomy of seed plants. Wiley, New York.

————. 1965. Vascular differentiation in plants. Holt, New York.

Laetsch, W. M., and R. E. Cleland [eds.]. 1967. Papers on plant growth and development. Little, Brown, Boston.

Maheshwari, P. 1950. An Introduction to the embryology of angisoperms. McGraw-Hill, New York.

O'Brien, T. P., and M. E. McCully. 1969. Plant structure and development. A pictorial and physiological approach. Macmillan, New York.

Romberger, J. A. 1963. Meristems, growth, and development in woody plants. U.S. Dept. of Agric., Forest Service, Technical Bull. **1293**:1–214.

Ruhland, W. [ed.]. 1965. Encyclopedia of plant physiology. Vols. XV/1 and XV/2. Differentiation and development. Springer-Verlag, Berlin.

Sinnott, E. W. 1960. Plant morphogenesis. McGraw-Hill, New York.

Steward, F. C. 1968. Growth and organization in plants. Addison-Wesley, Reading, Mass.

Torrey, J. G. 1967. Development in flowering plants. Macmillan, New York.

Wardlaw, C. W. 1955. Embryogenesis in plants. Wiley, New York.

———. 1968. Morphogenesis in plants. A contemporary study. Methuen, London.

Wareing, P. F., and I. D. J. Phillips. 1970. The control of growth and differentiation in plants. Pergamon, Oxford.

Wetmore, R. H., and C. W. Wardlaw. 1951. Experimental morphogenesis in vascular plants. Ann. Rev. Plant Physiol. **2**:269–292.

White, P. R., and A. R. Grove, [eds.]. 1965. Proceedings of an international conference on plant tissue culture. McCutchan Public Corp., Berkeley, Calif.

TWO

Embryogenesis—Beginnings
of Development

We have seen in the previous chapter that it is characteristic for morphogenetic events to continue throughout the life-span of most plants. This is in marked contrast to animal development, in which there is a concentration of morphogenetic phenomena in the embryonic stages. Nonetheless, the vascular plant, like the animal, begins its life as a single cell, the fertilized egg, and passes through an embryonic phase during which the fundamental body plan is laid down. Although it may be argued that all plants that develop from a single cell into a multicellular state pass through an embryonic phase, historically the term *embryo* has been restricted to those groups in which the early stages are enclosed within parental tissue and are presumed to be nutritionally dependent upon the parent organism. On this basis, the bryophytes and the vascular plants often are designated the *Embryophyta*. In the bryophytes and the lower vascular plants there is no interruption of growth to mark the end of the embryonic phase, which is, therefore, rather ill-defined. On the other hand, in the seed plants, embryonic development is considered to be terminated at the onset of seed dormancy; and this leads to a sharp distinction between the embryo and all postgermination stages.

Throughout the Embryophyta, as well as in some lower groups, embryo development from a zygote alternates in the life cycle with development of a second plant body from a single-celled spore. Alternation of generations poses interesting morphogenetic problems because of the contrasting morphology of the two phases, each developed from a single cell but under different conditions. In discussing embryogenesis we shall be concerned primarily with the early or embryonic stages of development of the diploid zygote into a vascular sporophyte. The contrasting development of the haploid spore into the gametophyte will not be considered except as it sheds some light upon the factors that control the different development of the two generations.

Patterns of embryo development

In view of the fundamental similarity of somatic organization in the sporophytes of principal groups of vascular plants, it is startling to discover the diversity of embryological patterns that lead to this organization. The conclusion is unavoidable that these patterns in themselves have limited morphogenetic significance. Nevertheless, the various patterns must be recognized, because any functional generalizations must be compatible with them. The classification of embryo types based on the sequence of early cell divisions is a complex field with a voluminous literature, and is one that has had an important bearing upon taxonomic and phylogenetic interpretations.

In this chapter we shall not examine examples of all these types, but rather we shall describe some that reveal the range of embryological diversity, with the hope of arriving at some general understanding of the principles of embryogenesis.

Embryo development in angiosperms

An excellent account of embryo development in a flowering plant has been given by Miller and Wetmore (1945) for *Phlox drummondii* (Figs. 2.1, 2.2). The first cell division in the zygote is at right angles to the axis of the embryo sac, in common with all angiosperms that have been studied. The two cells each divide again in the same plane giving rise to a four-celled filament. Divisions continue in each of these cells. The basal cell, which lies nearest to the micropyle—the aperture through which the pollen tube grows prior to fertilization—divides in the same plane, producing a short, filamentous organ called the *suspensor*. The remaining three cells undergo both longitudinal and transverse divisions and give rise to a globular mass in which the cells are arranged in regular tiers. At this stage the embryo, consisting of fewer than 40 cells, is only four days old and is less than a quarter of a millimeter in length. By the fifth day the first evidence of histodifferentiation within the previously homogeneous globular part of the embryo is detected. Divisions in the surface cells become progressively restricted to the anticlinal plane—

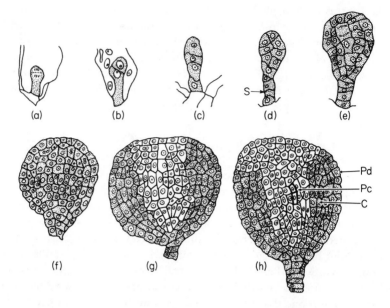

Fig. 2.1   Embryo development in *Phlox drummondii.* (a) First division of the zygote. (b–e) Stages of embryogeny in the first four days after fertilization. (f–h) Later stages showing differentiation of protoderm, procambium, and cortical parenchyma. The shoot apex is first distinguishable in (g). Key: C, cortical parenchyma; Pc, procambium; Pd, protoderm; S, suspensor. ×190. (H. A. Miller and R. H. Wetmore. 1945. Amer. J. Bot. **32**: 588.)

that is, perpendicular to the surface—resulting in the appearance of a superficial layer called the *protoderm.* Shortly thereafter, internal differentiation becomes evident in a central column of densely staining, narrow, elongated procambial cells surrounded by a cylinder of vacuolated cells. Thus, at this early stage, the principal tissue systems of the plant have been initiated.

By the sixth day, the first suggestion of the shoot apical meristem can be detected in the spherical embryo as an area of small, densely staining cells continuous with the central procambial core and at the pole of the embryo opposite the suspensor. A day or two later, the two cotyledons appear as a result of localized concentrations of growth on either side of the shoot meristem but not in it, and the embryo passes into what has been called the heart-shaped stage. Procambium continuous with that of the central core of the embryo axis extends into the cotyledons. As the cotyledons enlarge, the axis of the embryo also elongates and, at the end of the axis opposite the shoot apical meristem, periclinal divisions—that is in a plane parallel to the surface—initiate a root cap beneath which the apical meristem of the primary root can be detected. The mature

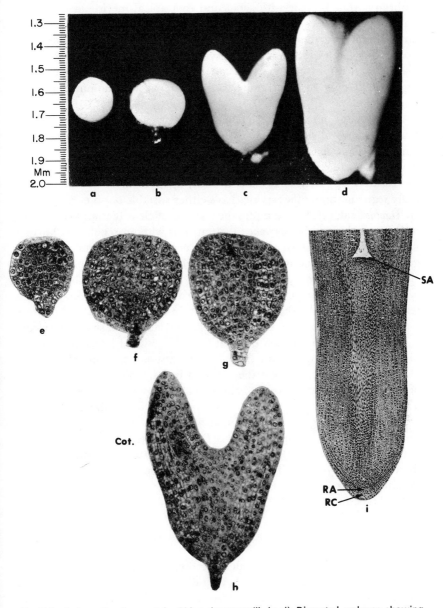

Fig. 2.2   Embryo development in *Phlox drummondii*. (a–d) Dissected embryos showing globular, heart, and early torpedo stages. (e–h) Sections of embryos showing cellular detail of comparable stages. (e), (f), and (g) correspond to stages (f), (g), and (h) in Fig. 2.1. (i) Fully developed embryo showing shoot and root apices. Key: Cot, cotyledon; RA, root apex; RC, root cap; SA, shoot apex. (a–d) ×61, (e–i) ×149. (H. A. Miller and R. H. Wetmore. 1945. Amer. J. Bot. **32**: 588.)

embryo is thus bipolar with shoot and root apical meristems located at opposite extremities of its axis. The mature embryo, like the full-term animal embryo, possesses the fundamental organization of the adult body; but unlike the animal embryo, in which the major organs are present at least in rudimentary form, the shoot and root systems are represented only by their respective meristems with unlimited growth potential.

On the basis of the planes of early cell divisions and the contributions that their several derivatives make to the development of the globular embryo, plant embryologists have recognized six types of embryonic development in the angiosperms. Over the years, however, it has become increasingly apparent that though many species have fixed patterns of early segmentation, other species have either variable or obscure patterns. In *Daucus* (carrot), for example, the pattern of cleavage has been shown to be variable, and in *Gossypium* (cotton) it is without regularity. Further variations in the pattern of development occur in later stages of embryogeny. Embryos of the monocotyledons, which up to the globular stage closely resemble embryos of dicotyledons and conform to the same six types, develop a single prominent cotyledon in what appears to be a terminal position, the shoot apex apparently occupying a lateral position on the embryo. There is uncertainty as to whether the single cotyledon is truly terminal in position or whether it is initiated laterally as in the dicotyledons and by growth displacement assumes a terminal position. Supporting the latter view is the fact that in some dicotyledons only one of the two cotyledons develops, and this shows a terminal displacement. There appear also to be differences in the extent of cellular proliferation in the globular stage before the onset of histogenesis and organogenesis. For example, in the primitive genus *Degeneria*, a tropical tree, there is a massive globular stage that precedes any differentiation. By contrast, in the rush *Luzula forsteri* differentiation is precocious, the protoderm being initiated when the globular part of the embryo consists of only eight cells (Maheshwari, 1950).

Embryo development in gymnosperms

Although there is considerable diversity in embryonic development among the gymnosperms, they all differ in characteristic ways from the angiosperms that have been considered. The embryology of *Ginkgo biloba*, an ancient and taxonomically isolated species, has been investigated extensively and will serve as a type for illustration (Fig. 2.3). As in all gymnosperms, the egg is large (300 to 500 microns in diameter in *Ginkgo*) and is contained within an archegonium, the female sex organ, at the micropylar end of a cellular female gametophyte. After fertiliza-

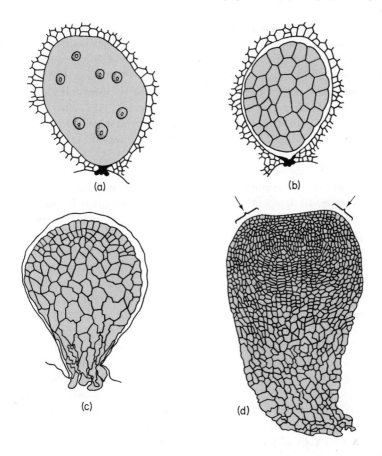

Fig. 2.3 Embryo development in *Ginkgo biloba*. (a) Early free nuclear embryo. (b) Wall formation complete throughout the embryo. (c) Cellular embryo at a later stage of development showing the axial gradient of cell size. (d) An early stage of cotyledon development. The arrows indicate the positions of the two cotyledons. (a, b) ×65, (c) ×75, (d) ×60. ((a, b) D. A. Johansen. 1950. Plant Embryol., Chronica Botanica, Waltham; (c, d) H. G. Lyon. 1904. Minn. Bot. Stud. **3**: 275.)

tion, the zygote nucleus divides but no cell wall arises to separate the daughter nuclei. These nuclei and their progeny divide repeatedly until about 256 nuclei are distributed uniformly throughout the cytoplasmic mass of the embryo. At this time, or after one further general division, wall formation occurs in such a way as to partition the embryo into uninucleate cells of equal size. The cells farthest from the micropyle divide more rapidly in the ensuing period than do those at the micropylar end; and the embryo at this time exhibits a marked axial gradient in cell size. These small cells, occupying about one third of the embryonic volume, give rise upon further development to the organized embryo with shoot and root apices and cotyledons. The larger cells, which make up the other two thirds of the embryo, divide more slowly, are vacuolated, and, according to some, constitute a suspensor (Wardlaw, 1955).

The phenomenon of free nuclear division at the beginning of embryo development is characteristic of the gymnosperms and occurs in no other plant embryos. The free nuclear stage may be even more extensive than in *Ginkgo*, as in the cycad *Dioon edule*, where over a thousand nuclei have been noted in the coenocytic embryo. Conversely this stage may be relatively brief, as in the conifers, in which the number of free nuclei may range from 32 or 64 in members of the family Araucariaceae to two in the Cupressaceae. In *Sequoia* the first division of the zygote is followed by cell wall formation, so that there is no free nuclear stage in the embryo of this species.

Another peculiarity of gymnosperm embryology is the widespread occurrence of multiple embryos arising from a single zygote. Cleavage polyembryony is best developed among the conifers, and may be illustrated in its most elaborate form in the genus *Pinus* (Fig. 2.4). The multiple embryos arise in characteristic positions, and in order to appreciate their development it is necessary to study the early stages of pine embryology. The four nuclei of the free nucleate stage move to the inner end of the embryo, opposite the micropyle, and there become arranged in a single tier. The nuclei then divide synchronously, producing a second tier of four nuclei, and this division is immediately followed by wall formation. The inner tier is completely enclosed by walls and the outer tier is open toward the micropyle. Each tier divides again so that there are now four tiers of four cells each, the outermost still open toward the micropyle. The cells of the innermost tier are called the apical cells, those of the second tier the primary suspensor cells, and those of the next tier the rosette cells. The suspensor cells begin to elongate, and their growth pushes the apical cells into the tissue of the female gametophyte. The apical cells separate laterally from one another, meanwhile cutting off secondary suspensor cells, which also elongate. Each apical cell, by dividing in various planes, builds up a multicellular mass that develops into a bipolar embryo. Even more striking is the fact that the rosette cells resume growth and, by a process similar to that occurring in the apical cell derivatives, each gives rise to a separate embryo. Thus up to eight embryos may result from the development of a single zygote (Wardlaw, 1955).

The later stages of embryogeny, which have been investigated in pine, in many ways resemble those of angiosperms. The first embryonic organ to appear is the root apex with a massive root cap adjacent to the suspensor. At an early stage histodifferentiation begins, as evidenced by the

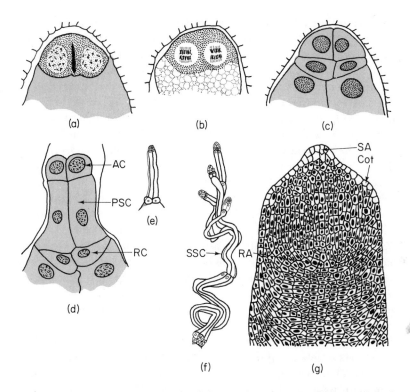

Fig. 2.4   Embryo development in *Pinus*. (a) Free nuclear stage showing two of the four nuclei at the end of the embryo opposite the micropyle. (b–d) Successive stages in the development of the four cell tiers. The suspensor cells are beginning to elongate in (d). (e, f) Stages in elongation of the suspensors and development of multiple embryos. (g) The embryo showing the differentiation of organs and tissues. Key: AC, apical cells; Cot, cotyledon; PSC, primary suspensor cells; RA, root apex; RC, rosette cells; SA, shoot apex; SSC, secondary suspensor cells. (a) ×90, (b–d) ×135, (g) ×95. ((a–d) D. A. Johansen. 1950. Plant Embryol., Chronica Botanica, Waltham; (e, f) J. T. Buchholz. 1931. Trans. Ill. Acad. Sci. **23**: 117; (g) A. R. Spurr. 1949. Amer. J. Bot. **36**: 629.)

appearance of vacuolated pith parenchyma and elongated procambial cells. Finally the shoot apex arises and gives rise to the multiple cotyledons (Spurr, 1949). Thus an embryo develops much as in *Phlox*, but the sequence of morphogenetic events is rather different.

Embryo development in lower vascular plants

In the lower vascular plants there is a bewildering array of embryonic types, which have been described in more or less detail by numerous authors. In the majority of ferns, the gametophytic phase of the life cycle is a distinct, free-living organism of diminutive size dependent upon its own photosynthesis for the energy required not only for its own development but for that of the embryo as well. Under these conditions one might expect to find the pattern of embryonic development in these organisms very different from those patterns found in the well-nourished embryos of the seed plants. In fact, the general pattern of embryogeny is not markedly dissimilar from embryogenic patterns of higher plants. Much of the classical work on fern embryology has been concerned with early cleavage patterns and the possible organogenetic significance of this segmentation. In the leptosporangiate ferns (Fig. 2.5), with few exceptions the first division of the zygote is parallel to the long axis of the archegonium. The second division is perpendicular to the first and usually at right angles to the archegonial axis, but sometimes, for example in *Todea*, parallel to it. In either case, the four quadrants divide again, usually synchronously, to form the octant stage embryo. The subsequent development of organs has been related to the original quadrants, each giving rise to a single organ, the shoot, the first leaf, the foot (an embryonic organ having a presumed absorptive function), and the root. The first division beyond the octant stage was presumed to set off the apical cells of the root, shoot, and first leaf.

The precise relationship between the quadrants and the embryonic organs has often been questioned, and several investigations have thrown doubt upon its general applicability. Ward (1954) has reported that in the embryogeny of *Phlebodium aureum* (Fig. 2.5) organs become recognizable only after many cell divisions and much growth beyond the quadrant stage. Cell divisions beyond the octant stage produce a globular mass within which differentiation begins, and the first evidence of this is in the part of the embryo that will give rise to the foot. The foot appears to originate from much of the tissue derived from the two upper quadrants of the early embryo and not from one of these alone as the classical interpretation had suggested. Development of the foot involves the enlargement and vacuolation of cells that remain thin-walled and intrude between some of the adjacent cells of the gametophyte. Soon thereafter the first leaf appears as an outgrowth from the region referable to the lower anterior quadrant, in accordance with the classical description. The root arises next from derivatives of the lower posterior quadrant; and its origin, like that of the leaf, appears to conform to the traditional interpretation. The final embryonic structure to develop is the shoot apex, which appears between the leaf and the foot and is not referable to an entire quadrant. The four embryonic organs appear at about one-day intervals. Whereas classical fern embryology implied that organs are delimited at a very early stage in relation to segmentation of the zygote,

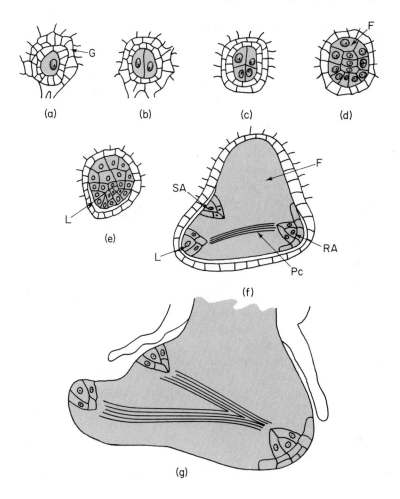

Fig. 2.5   Embryo development in *Phlebodium aureum*. (a–c) Early developmental stages in which cell division in the embryo proceeds without evident differentiation. (d) Cells in the upper half of the embryo enlarge to form the foot. (e) Origin of the first leaf in the embryo. (f, g) Later stages of embryo development showing the origin and growth of the shoot and root apices, and the position of the procambium. Key: F, foot; G, gametophyte cells; L, leaf; Pc, procambium; RA, root apex; SA, shoot apex. (a–f) ×150, (g) ×120. (Drawn from M. Ward. 1954. Phytomorphology **4**: 18.)

this study and others like it show the prior formation of a multicellular mass within which the organs are delimited sequentially.

Some additional features of fern embryogeny have been revealed in a more recent study, that of DeMaggio (1961) on *Todea barbara*. In this fern, the first evidence of differentiation in the embryo is the enlargement and vacuolation of cells in the foot region, as was the case in *Phlebodium*. In the remainder of the embryo, differentiation into an outer, epidermal-like layer and a central group of very small cells then takes place. It is only after this differentiation that the other embryonic organs have their origins. The leaf is the next organ to appear, but unlike the sequence in *Phlebodium*, the shoot apex arises before the root.

The practice of interpreting embryogenesis in relation to early cleavage patterns has been characteristic of studies of other groups of lower vascular plants as well, and in the absence of modern reinvestigations, one questions this type of interpretation on the basis of comparison with the ferns. In both *Lycopodium* and *Selaginella* (Lycopsida) the first division of the zygote is transverse and cuts off a cell adjacent to the archegonial neck, which develops into a suspensor of limited extent. In *Lycopodium* the suspensor cell only rarely divides; but in *Selaginella* it often becomes multicellular. A foot ordinarily develops in the region of the embryo adjacent to the suspensor, and it is usually massive in *Lycopodium* but variable in extent of development in *Selaginella*. Another interesting feature of both of these genera is the delayed development of an embryonic root that appears long after the shoot apex and first leaf are well developed. The delay in root initiation in the primitive Lycopsida is of interest in relation to the complete absence of roots from both the embryo and the adult plant of both living members of the Psilopsida (*Psilotum* and *Tmesipteris*). In all of the lower vascular plants, it will be recalled, there is no interruption of growth to mark the end of a distinct embryonic phase. The embryo develops directly into a juvenile plant, which bursts out of the enclosing parental gametophytic tissues and in a relatively short time becomes independent. In some species of *Lycopodium*, the embryo ruptures the surrounding gametophytic tissues before differentiation of embryonic organs and grows out as a green parenchymatous structure, the protocorm, on which a shoot apex is subsequently differentiated.

Cellular changes during embryogenesis

Fertilization marks the onset of extensive change and reorganization of the egg cell cytoplasm, and this has been described by Jensen (1968) in a recent study of *Gossypium* (cotton) and by Schulz and Jensen (1968) in *Capsella* (shepherd's purse) (Fig. 2.6). There are marked changes of appearance and distribution of cell organelles in the zygote. Ribosomes become aggregated into long helical polysomes which are associated with plastids and mitochondria. Later new ribosomes begin to be synthesized but these aggregate into shorter polysomes which can be distinguished from those formed from maternal ribosomes on the

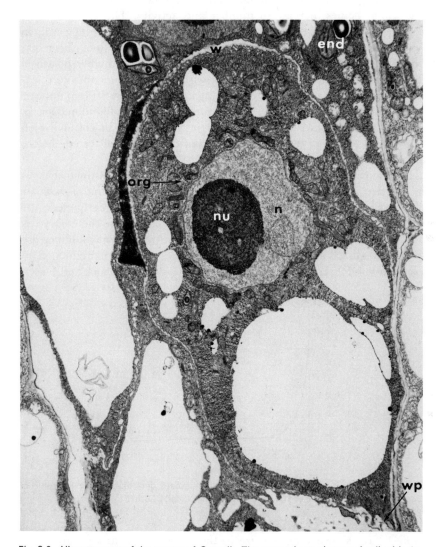

Fig. 2.6   Ultrastructure of the zygote of *Capsella*. The zygote is an elongated cell with the nucleus situated at the terminal end and a large vacuole at the basal end. Other smaller vacuoles occur throughout the cytoplasm. Plastids and mitochondria are concentrated around the nucleus. The cell wall separating the zygote from the surrounding endosperm is becoming thick, and at the basal end of the zygote there are wall projections that increase the absorbing surface. Key: end, endosperm; n, nucleus; nu, nucleolus; org, cytoplasmic organelles concentrated near the nucleus; w, wall of the zygote; wp, wall projections. ×5520. (R. Schulz and W. A. Jensen. 1968. Amer. J. Bot. **55**: 807.)

basis of length. Plastids and mitochondria accumulate around the nucleus, and Golgi bodies increase in number and activity. The effect of these changes is to convert the seemingly metabolically inactive egg cell, in which organelles are arranged more or less randomly, into the metabolically active zygote cell in which organelles have a highly polarized distribution. In addition, in *Gossypium*, within a few hours after fertilization a distinctive type of tube-containing endoplasmic reticulum appears in the cytoplasm of the zygote. This feature, which is not found in *Capsella*, may be associated with the reduction in vacuolation in cotton which causes a shrinkage of the zygote to one-half the volume of the egg before any divisions occur.

During early stages of embryo development there are distinct changes in cell size which appear consistently in the various groups of vascular plants. The zygote is, characteristically, a large cell. The early cleavages of the embryo result in successively smaller cells because the embryo as a whole either increases slowly in size during the process, may not increase in size at all, or sometimes actually decreases because of reduction in vacuolar volume. In the embryo of *Gossypium* (Fig. 2.7), Pollock and Jensen

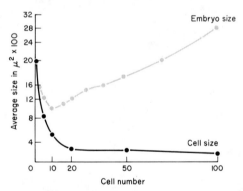

Fig. 2.7   Changes in cell size and embryo size during early development of the embryo of *Gossypium*. (E. G. Pollock and W. A. Jensen. 1964. Amer. J. Bot. **51** : 915.)

(1964) have shown that cell size in the embryo decreases progressively until, at the 100-cell stage, average cell size is about one twentieth that of the zygote. Cell size then remains relatively constant until about the 1000-cell stage when it is further reduced by one half. The reduction in cell size continues for a variable length of time in embryos of different species, and may then remain constant during further growth until changes associated with histodifferentiation appear. Comparable reductions of cell size are a well-known feature of cleavage in animal embryos in which it has been suggested that the significance of size reduction is to establish a desirable balance between the size of the nucleus and the volume of the associated cytoplasm. No doubt the same value may be assigned to this phenomenon in plants. In addition, it would seem that the onset of histogenesis and organogenesis in an embryonic mass of small size would require that the protoplasm be compartmentalized, and that it may be essential to increase the number of cells rapidly with a minimum increase of total mass.

In all vascular plant groups there is an early distinction established between cells that remain small and inconspicuously vacuolated and those that enlarge and become conspicuously vacuolated. Cells of the former region become the principal histogenetic and organogenetic portions of the embryo and those of the latter region have been assigned a variety of functions. The vacuolated regions in different groups have diverse origins and structures and consequently have been given various names such as suspensor or foot. The universal occurrence of such regions suggests that they have developmental significance, and one possible role is that of absorbing nutrients from the surrounding tissues and transmitting them to the organogenetic part of the embryo. Although such function has received only scant attention in the seed plants, this absorption long has been considered to be the function of the foot in lower vascular plants.

General comment

Several general observations emerge from a comparative examination of embryonic development in the vascular plants. In spite of apparent great diversity, there is a strong suggestion of fundamental similarity of pattern in the establishment of the basic developmental plan of the embryo. The obvious differences that occur are those relating to early cleavage patterns before the onset of histogenesis or organogenesis, those concerned with accessory structures pertaining to embryonic survival, and those resulting from differential growth rates among the various embryonic tissues and organs. Thus, though the free nuclear stage of the gymnosperm embryo contrasts sharply with the cleavage patterns of other groups, though the absence of a suspensor in leptosporangiate ferns contrasts with its constant presence in the lycopods, and though the precocious initiation and enlargement of the first leaf in fern embryos is a distinctive and consistent feature, these features are somewhat peripheral to the basic development of the embryo though each has its own morphogenetic interest. Thus, diverse patterns of embryogeny seem to converge upon a common point, the production of an embryo having the rudiments of the shoot and root systems.

From the morphogenetic point of view, one of the most interesting conclusions arising from the study of embryonic development is that the developmental relationships of organs and tissues to meristems that one has come to recognize as general from the study of adult plant development do not necessarily hold for embryonic stages. In the adult plant, leaf primordia arise only in relation to the activity of the shoot apical meristem, but in many embryos the emergence of embryonic leaves may be unrelated to a localized meristem. Similarly, there are numerous instances in which histodifferentiation in embryos is initiated independently of any organized meristem of root, stem, or leaf, whereas in the adult plant the meristem commonly has been regarded as the initiator and organizer of these events. This leads to the not surprising conclusion that the regulation of morphogenetic processes at the time of their inception in the embryo is rather different from the regulation of their continuation in the expanding plant. It also calls attention to the possibility that the environment in which the embryo develops has an important bearing upon the control of morphogenetic processes. It is to these questions that attention will be turned in the next chapter.

REFERENCES

DeMaggio, A. E. 1961. Morphogenetic studies on the fern *Todea barbara*. (L.) Moore. II. Development of the embryo. Phytomorphology **11**:64–79.

Jensen, W. A. 1968. Cotton embryogenesis: the zygote. Planta. **79**:346–366.

Maheshwari, P. 1950. An introduction to the embryology of angiosperms. McGraw-Hill, New York.

Miller, H. A., and R. H. Wetmore. 1945. Studies in the developmental anatomy of *Phlox drummondii* Hook. I. The embryo. Am. J. Botany **32**:588–599.

Pollock, E. G., and W. A. Jensen. 1964. Cell development during early embryogenesis in *Capsella* and *Gossypium*. Am. J. Botany **51**:915–921.

Schulz, Sister R., and Jensen, W. A. 1968. *Capsella* embryogenesis: the egg, zygote and young embryo. Am. J. Botany **55**:807–819.

Spurr, A. R. 1949. Histogenesis and organization of the embryo in *Pinus strobus*. L. Am. J. Botany **36**:629–641.

Ward, M. 1954. The development of the embryo of *Phlebodium aureum*. J. Sm. Phytomorphology **4**:18–26.

Wardlaw, C. W. 1955. Embryogenesis in plants. Wiley, New York.

THREE

Experimental and Analytical Studies of Embryogenesis

The particular conditions under which development begins in the embryo of vascular plants obviously hold great interest for the student of morphogenesis. The question that clearly requires investigation is whether the particular pattern that emerges during embryogeny is to be regarded as an expression of the inherent capacity of the zygote, as the result of specific regulation from the environment, or as the manifestation of subtle interaction between the two. Although descriptive accounts of embryogenesis are helpful in exploring this problem, it is to experimental and analytical techniques that one must turn for further exploration of these possibilities. Experimental embryology has been an extremely valuable discipline in elucidating problems of animal morphogenesis; but the plant counterpart of this field has played a limited role in the understanding of plant morphogenesis. A major factor contributing to this deficiency has been the relative inaccessibility of the plant embryo at the formative stages, with the result that the botanical work that most closely corresponds to experimental animal embryology has been done with the apical meristems of the adult plant. Nonetheless, there have been several pioneering studies in the field of experimental plant embryology that bear upon this problem, and it will be useful to examine these now.

The external influences to which the zygote and young embryo are exposed are complex and interrelated, but for the purpose of analysis they may be treated in three categories: the fertilization stimulus, chem-

21

ical factors in the embryonic environment, and physical phenomena that influence development.

Fertilization and egg activation

It is generally recognized in animal embryology that the process of fertilization plays two important roles in the initiation of embryogenesis, and these have been discussed at length by Austin in another book in this series.* There is the obvious union of two haploid gametic nuclei, which establishes the diploid condition of the zygote. In addition, fertilization provides an activation stimulus, which causes the zygote to begin cleavage. That these are two separate phenomena has been established experimentally by substituting artificial activators for the sperm. Botanists, on the other hand, have been less concerned with the activation stimulus. There are, however, a number of well-documented cases in which the stimulus to egg development is derived from some source other than fertilization. Maheshwari (1950), though concluding that there is no reliable treatment for the induction of parthenogenesis in higher plants, lists four that have been successful in specific instances. These are exposure to high or low extremes of temperature soon after pollination, pollination by X-ray inactivated pollen or that of another species, delayed pollination, and chemical treatments. These cases of artificial activation of the egg, although not common, clearly show that the stimulus is distinct from the nuclear aspects of fertilization. In this connection it is interesting to note that ultrastructural studies in cotton (*Gossypium hirsutum*) have shown that only the sperm nucleus enters the egg in the process of fertilization (Jensen and Fisher, 1967). Moreover, it seems likely that there are many unrecognized instances of parthenogenesis, in the angiosperms at least, in which the embryo dies after a few cleavages owing to the failure of development of the nutritive endosperm that is also dependent upon fertilization. The fact that a variety of seemingly unrelated stimuli are effective as activators of development strongly suggests that fertilization stimulation acts as a trigger mechanism, setting in motion a sequence of events that is not dependent upon the precise nature of the stimulus. The nature of the true fertilization stimulus is not revealed by any of the observations reviewed here; and there are additional observations that must be considered also. In some cases after fertilization and the beginning of growth in the diploid embryo, one of the accessory synergid cells adjacent to the embryo is stimulated to develop without fertilization and both embryos continue to develop, producing a seed with twin embryos, one diploid and the other haploid. Similarly, there are other cases in

* Austin, C. R. 1965. Fertilization. Prentice-Hall, Englewood Cliffs, N.J.

which, following fertilization, additional embryos arise from diploid cells surrounding the embryo sac. In these cases of adventive embryony, it is apparent that some stimulus derived from the original fertilization must be passed to the other cells, because in the absence of fertilization, accessory embryos are not initiated. It is surprising in these cases that the activation stimulus arising from the developing embryo should have an effect upon so few of the surrounding cells in the ovule and that in normal embryogeny, no cells other than the zygote are activated. Considerations such as these raise interesting questions as to whether the egg acquires during maturation characteristics that cause it to respond uniquely to the stimulus of fertilization. This question will be considered later in this chapter.

The environment of the embryo

The morphological environment

Whether or not the egg is a unique cell, it is evident that the environment under which it and later the zygote develop is strikingly different from that in which postembryonic development occurs. The embryo is surrounded by a tissue from which it derives nutrients during its development. Although attention has been focused largely upon the nutritional aspects of the embryonic environment, it is becoming increasingly clear that development is influenced by a variety of hormonal stimuli that arise in the surrounding tissue.

The tissues that surround the developing embryo in the various groups of vascular plants are remarkably diverse in structure, origin, physiological characteristics, and physical relationship to the embryo. The relationship between the embryo and its surroundings is most complex, but perhaps most familiar, in the flowering plants (Fig. 3.1). Prior to fertilization, the egg is attached at the micropylar end of the embryo sac and extends, with its two synergids, freely into the embryo sac cytoplasm. Though there is evidence that the embryo sac contains nutritive substances that are utilized in the early cleavage stages of the zygote, it is clear that most of embryonic development depends upon nutrients that accumulate in a specialized tissue called the endosperm. Like the zygote, the endosperm results from events occurring at fertizilation and it develops simultaneously with the embryo. The endosperm is initiated by a fusion between the second sperm and a diploid nucleus, which has itself arisen from the fusion of two haploid nuclei of the embryo sac. There are various patterns of endosperm development. In some cases nuclear division is accompanied by wall formation so that the tissue is cellular from the

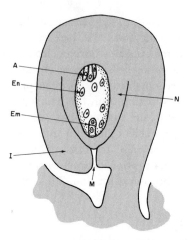

Fig. 3.1 Longitudinal section of an angiosperm ovule showing details of the embryo sac shortly after fertilization. Key: A, antipodal cells; Em, embryo after first division; En, endosperm at free nuclear stage; I, integuments; M, micropyle; N, nucellus.

beginning. In other cases, a free nuclear phase precedes wall formation. This phase may be of limited duration or, as in coconut and other species with so-called liquid endosperm, may be prolonged. The customary dependence of the embryo upon the endosperm for a continuing supply of nutrients is well demonstrated by those instances of interspecific crosses in which the embryo begins to develop but aborts as a result of failure of endosperm development.

The embryo is thus provided, throughout its development, with a nurse tissue that accumulates materials brought into the ovule from other parts of the sporophyte and retains them in a form usable by the embryo. Although it has long been assumed that the embryo suspensor has no special function other than that of pushing the organogenetic part of the embryo into the developing endosperm, it seems highly probable that this embryonic organ has a more direct role in the nutrition of the embryo. The vacuolating cells of the suspensor may be a preferred path of entry for nutrients into the more meristematic regions of the embryo. The elaborate haustorial development of the suspensor in some species provides support for this view (Fig. 3.2), as does the formation of internal wall projections which have the effect of increasing substantially the surface area of the plasma membrane in suspensor cells (Gunning and Pate, 1968). There are instances in which the haustorial suspensor actually penetrates the sporophytic tissues of the ovule or even tissues beyond the ovule, thus bypassing completely the endosperm nutritional mechanism (Maheshwari, 1950).

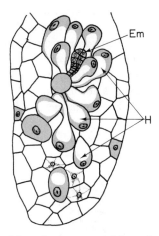

Fig. 3.2 Haustoria developed from the suspensor of *Asperula*. Key: Em, main body of the embryo; H, haustorial suspensor. (F. E. Lloyd. 1902. Mem. Torrey Bot. Club **8**: 1.)

As in the angiosperms, the developing gymnosperm embryo is surrounded by a tissue rich in nutrients, but the nutritive tissue in the gymnosperms is the haploid female gametophyte, and it is developed and well stocked with nutrients before fertilization occurs (Fig. 3.3). The embryo initially is located at the micropylar end of the female gametophyte and during its development it is progressively forced into the surrounding tissue by growth of the suspensor. It seems possible that

Fig. 3.3 Longitudinal section of a gymnosperm ovule (a cycad). Key: Ar, archegonium; Em, two-nucleate stage embryo; F, female gametophyte; I, integument; M, micropyle; N, nucellus. ×3.

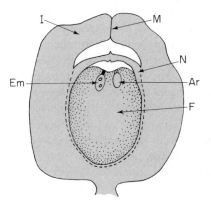

here, as in the angiosperms, the vacuolating cells of the suspensor may play a direct role in the absorption of nutrients. If this is true, the elaborate development of the suspensor in some groups might be a physiological parallel to the elaboration of this embryonic organ in some angiosperms. In both groups of seed plants, the absorption of nutrients from the surrounding tissue by the embryo is accompanied by a disintegration of the nutritive tissue, and the embryo enlarges to fill the space previously occupied by these cells.

In the ferns (Fig. 3.4), the physical and physiological relationships

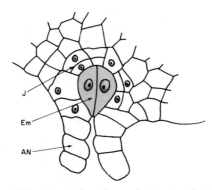

Fig. 3.4   The embryo and the surrounding tissues in the fern *Todea barbara*. Key: AN, archegonial neck; Em, two-celled embryo; J, jacket of gametophyte tissue. ×250. (Drawn from A. E. DeMaggio. 1961. Phytomorphology **11** : 46.)

between the developing embryo and the tissues that provide its nourishment are quite different. This nourishment is provided by the diminutive gametophyte and, in contrast to the nutrients available to embryos of seed plants, it appears to be limited in quantity and perhaps simple in composition. The embryo develops an intimate contact with the tissues of the gametophyte, cells of the foot often interdigitating with those of the gametophyte, and cells of both sporophyte foot and gametophyte developing internal wall projections like those of the angiosperm suspensor (Gunning and Pate, 1968). Moreover, the cells of the gametophyte that are adjacent to the embryo respond to its presence by undergoing a series of periclinal divisions to form a multilayered jacket. This has been interpreted as placing restraint upon the expansion of the embryo as it grows and it is regarded as having morphogenetic significance. The gametophyte in other groups of lower vascular plants differs from that of the ferns either by being saprophytic or by developing largely within the spore and being dependent upon food reserves contained within the spore.

The chemical environment

In attempting to understand the role of nutrients and growth-promoting substances in the development of the embryo, many workers have utilized the technique of embryo culture in which the embryo is removed from its natural environment at the desired stage of development and placed in a nutrient medium of known composition (Raghavan, 1966). The use of this technique has a long history, extending back into the last century. During the early period a number of workers removed fully developed or nearly mature embryos and found that these were able to germinate satisfactorily on nutrient media of very simple composition. In some cases the nutrients consisted of solutions of inorganic salts only. Although the culture of mature embryos has been of great significance to horticulturalists concerned with overcoming dormancy problems in some seeds, it tells little about the morphogenetic control of embryogeny. In order to approach problems in this area it is necessary to culture embryos at earlier stages of development. From the earliest attempts to excise and culture immature embryos, it has been apparent that they can be grown successfully only when the medium contains additives more complex than those required for mature embryos, and that progressively younger embryos have more exacting requirements. These results generally have been interpreted as revealing a progressive transition from a heterotrophic mode of nutrition towards autotrophy.

The literature dealing with the culture of young embryos shows that the basis for such a conclusion is far from well established. This is illustrated by examination of some studies in which progressively younger embryos have been grown successfully in culture. Early work on embryo culture of *Datura* (Van Overbeek et al., 1942) showed that mature embryos, or even well-differentiated immature embryos, could be grown into seedlings on a medium consisting only of inorganic salts and a low concentration of sugar. Less mature embryos failed to develop satisfactorily on this medium but could be cultured at the so-called torpedo stage— when the embryo is cylindrical with all organs formed but not fully enlarged—if the medium was supplemented by a mixture of organic substances including vitamins, amino acids, and other compounds believed to have growth-promoting activity. Still younger embryos, however, failed to develop even on this enriched medium. Embryos at the heart stage could be grown successfully only if the medium was further supplemented by coconut milk, a liquid endosperm selected because of its natural role in embryonic nutrition. A significant feature of the growth of these embryos was that for about ten days they developed in an embryonic way rather than undergoing changes that lead to germination and development of seedlings. Rapid cell elongation associated with precocious germination had characterized the growth of larger embryos cultured without coconut milk. From these results it was concluded that coconut milk contains one or more growth factors, collectively designated *embryo factor*, that promote the growth of young embryos without germination. Results of this type have been interpreted as revealing the progressive development in the embryo of the capacity to synthesize essential substances.

Other substances have been found that replace coconut milk in supporting the development of young embryos of *Datura* in culture, and that are, therefore, presumed to contain embryo factor. Among these are yeast extract, wheat germ, almond meal, and sterile-filtered malt extract. Subsequently it was learned that a mixture of amino acids in the form of casein hydrolysate will essentially replace the malt extract component of the culture medium. In work with isolated *Hordeum* embryos (Zieber et al., 1950), it was found that casein hydrolysate stimulates embryonic development as opposed to precocious germination. By testing the amino acid, inorganic phosphate, and sodium chloride components of commercial casein hydrolysate separately they found that much of the effect of casein hydrolysate was attributable to the increased osmotic pressure produced in the medium, although the amino acids also played a nutritional role.

The significance of high osmotic pressure in the culture of young embryos has intruded into the consideration of nutritional and hormonal aspects of embryo growth on a number of other occasions. It was found in *Datura* (Rietsema et al., 1953) that embryos excised at different developmental stages require different minimal sucrose concentrations for growth: globular, preheart stages, 8–12 percent; heart stages, 4 percent; and later stages require progressively lower concentrations until mature embryos can be grown in the complete absence of sugar. Interestingly enough, at a uniform concentration of 2 percent sucrose, embryos of different stages respond in exactly the same way to varied concentrations of mannitol, which suggests that the total effect of increased sucrose concentration is osmotic, the youngest embryos growing best in the medium with the highest osmotic pressure. Rijven (1952) working with *Capsella* embryos showed essentially the same relationship and demonstrated also that very small embryos are isotonic with higher osmotic values than are more advanced embryos. Rijven's experiments clearly showed that germination or expansion of embryos in culture occurred only after transfer to a medium with low osmotic value, thus strongly suggesting the role of high osmotic pressure in maintaining embryonic development.

The dependence of continued embryonic development of immature excised embryos in culture upon complex substances that might be considered to be of the embryo factor type appeared to be clear from early studies; but subsequent analysis has, as has been seen, greatly reduced the clarity of the picture. A later study by Raghavan and Torrey (1963) on globular embryos of *Capsella* has brought the picture into much clearer focus (Fig. 3.5). These workers found that relatively mature embryos

Fig. 3.5   Embryo culture of *Capsella bursa-pastoris*. (a) An early heart stage embryo dissected from the ovule. The part above the suspensor is 80 μm in length. (b) An embryo that was grown in culture for five weeks in darkness. The meristems of the shoot and the root are indicated by arrows. (c) An embryo that was grown in culture for five weeks in the light. A pair of leaves has been formed at the shoot apex. (a) ×62, (b, c) ×73. (V. Raghavan and J. G. Torrey. 1963. Amer. J. Bot. **50**: 540.)

could be cultured satisfactorily on simple media consisting of vitamins, inorganic salts, and a 2 percent concentration of sugar. Embryos as young as the heart stage grew satisfactorily, if somewhat slowly, on this medium, but younger embryos did not. To obtain growth of globular stages it was necessary to supplement this medium with indoleacetic acid, kinetin, and adenine sulphate; and they found a sharp optimum for each of these substances. There appeared to be no requirement for a high osmotic pressure of the medium. However, the surprising fact was that the requirement for the additives was in large part replaced by raising the sugar concentration of the basal medium to 12 or 18 percent or the salt concentration by a factor of 10. Thus it appears that growth can be supported either by complex substances of the embryo factor type or by high osmotic concentrations in the surrounding medium, and the real basis of the maintenance of embryonic development in isolated embryos growing in culture remains obscure.

The physical environment

It is now necessary to give consideration to the possible role of physical factors in the embryonic environment. As the embryo develops it expands into regions previously occupied or simultaneously being occupied by other tissues; and the probability exists that at least in some cases the embryo experiences physical restraint on its expansion that might have morphogenetic significance. It is reasonable to expect that physical factors in the embryonic environment might be most easily analyzed in groups such as the ferns in which the nutritional relationship between the embryo and its surrounding tissues appears to be relatively uncomplicated, and the archegonium within which the embryo develops provides an obvious source of physical restraint. The first attempt to approach this problem experimentally in a fern (*Phlebodium aureum*) was undertaken by Ward and Wetmore (1954). In this investigation various patterns of vertical and horizontal incisions were made in the prothallial tissues immediately surrounding the zygote (Fig. 3.6). It was presumed that the incisions would reduce the restraint normally placed upon the zygote and early embryo by these tissues, but this was not verified by direct measurement. Following such operations the development of the embryo was markedly disturbed and the rate of growth was somewhat slower than in normal embryogeny. The embryos burst out of the surrounding tissues in fewer days after fertilization than in normal cases. At this time differentiation of leaf, shoot apex, and root had not occurred, and in many cases, the embryo consisted of a cylindrical mass of tissue that subsequently produced a leaf-like organ at its tip followed by a shoot apex in a position lateral to the leaf. Root development was considerably delayed. In those cases where the operation consisted only of the removal of the superficial part of the archegonium by a horizontal cut, the resulting outgrowth was tuberous and irregular rather than cylindrical. After considerable delay, leaf-like appendages followed by laterally placed shoot apices arose in several places so that, in effect, one embryo produced several sporophytic buds. These experimental procedures clearly disturbed the normal pattern of embryogeny, but it is significant that in all cases the usual sporophytic organs differentiated. It is probable that the developmental abnormalities resulted from the reduction of physical restraint upon the embryo, but it cannot be overlooked that incisions in the gametophyte must also have impeded the normal nutritional and hormonal patterns and have caused the release of substances from damaged cells.

The more tuberous of the outgrowths described by Ward and Wetmore are suggestive of the early protocorm stage of embryogeny characteristic of certain species of *Lycopodium*, specifically those whose gametophytes are green and protrude above the surface of the substrate so that the embryos are formed above ground. In these species the embryo breaks out of the archegonium at an early stage and develops as an irregular mass lacking sporophytic organs, which develop subsequently from it.

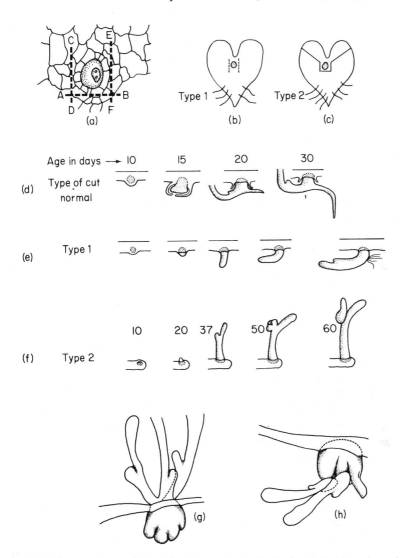

Fig. 3.6 Experiments on the effect of physical restraint on the development of the embryo of *Phlebodium aureum*. (a) Longitudinal section of the zygote and surrounding gametophytic tissues showing the pattern of vertical and horizontal incisions. An incision along the line A–B removes the archegonial neck. (b, c) Gametophytes showing the positions of vertical incisions that are designated Type 1 and 2 below. In both types the archegonial neck is also removed. (d) Development of an unoperated embryo as the control for experimental treatments. (e, f) Development of embryos after Type 1 and 2 operations. (g, h) Development of embryos after removal of the archegonial neck only. (Adapted from M. Ward and R. H. Wetmore. 1954. Amer. J. Bot. **41** : 428.)

In those species of *Lycopodium* with subterranean gametophytes, the embryo remains within the confining archegonium and a protocorm stage is lacking. Freeberg (1957) has made the interesting observation that in sterile culture, gametophytes that are normally subterranean assume a growth habit indistinguishable from that of the superficial species and produce a protocorm stage during embryogeny similar to that of the normally superficial species. These observations suggest that the tuberous protocorm results whenever the embryo is able to protrude at an early stage from the confining archegonium, an event which is normally impossible in the massive, subterranean gametophytes when these are growing under natural conditions, but which is highly probable in the filiform gametophytes of superficial species. This may offer an interesting parallel to the *Phlebodium* experiments, and one that is not subject to the same limitations.

It might be expected that the role of external physical constraint could be investigated most effectively if the whole course of embryonic development could be made to occur in its absence. This was done experimentally by DeMaggio and Wetmore (1961) when the zygote of *Todea* was removed from the archegonium and grown in aseptic culture. Growth of the cultured embryos was slow, but the planes of early divisions were oriented normally, producing octant-stage embryos. Later divisions were less regular than in the normal embryo, and division soon ceased. The cells then expanded, causing the surfaces to protrude in various ways. Development was arrested at this stage; but some months later the embryos were found to have resumed development and to have produced flattened, two-dimensional outgrowths reminiscent of gametophytic prothalli. Although this experiment is difficult to interpret fully, it provides striking confirmation of the speculation that the difference in growth pattern of the fern sporophyte and gametophyte may result from the different environments in which the development of each begins. The sporophyte begins as a physically constrained zygote, while the gametophyte begins as a freely exposed spore.

The role of physical factors in the control of embryogeny in flowering plants is less easily approached. The effect of confinement upon later stages of embryo development is evident in many cases in the curvature and other deformations that fit the embryo neatly into the seed, and that are notably lacking in embryos grown in culture. However, such effects of restriction in late stages of development are scarcely comparable to those suggested for the ferns that operate at very early stages, and they seem to be of little morphogenetic significance. It is precisely at the early stages of embryogeny in angiosperms that physical factors are least apparent, the early embryo being suspended freely in the fluid of the embryo sac. Only later as the endosperm develops does it appear to come into close physical contact with other tissues. The possibility exists that hydrostatic pressures such as those demonstrated in the female gametophytes of cycads might play a role in angiosperm embryogeny. Such pressures have been shown to be important in regulating cleavage planes in certain invertebrate embryos. In many cases in which very young embryos of flowering plants have been cultured, atypical developments such as supernumerary cotyledons, various outgrowths, and changed proportions between embryonic organs have been noted. In one instance (Norstog, 1961), the development of barley embryos excised at a very early stage was markedly altered and masses of tissue were produced on which several shoot and root apices were later initiated (Fig. 3.7). These structures are reminiscent of those produced in *Phlebodium* after the surgical removal of the archegonial neck. There is no clear evidence that any of

Fig. 3.7   Culture of barley embryos on a medium containing coconut milk. (a) Late stage embryo at the time of excision. (b, c) Development in culture of an embryo such as that shown in (a). Development has been relatively normal. (d) Early stage embryo at the time of excision. (e) Development in culture of an embryo such as that shown in (d). Development has been atypical and irregular. Key: C, coleoptile; L, leaf; R, root; S, shoot. (K. Norstog. 1961. Amer. J. Bot. **48**: 876.)

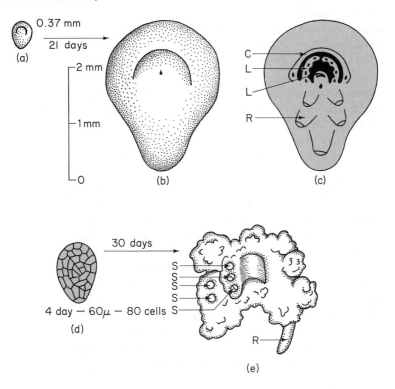

these abnormalities results from altered physical conditions in the environment of the embryo, but the possibility remains that physical factors may be at least partly responsible.

We have examined the diversity of embryological patterns and the numerous natural and experimentally induced conditions in which they can be elicited in the hope that the basic principles of embryogeny in the vascular plants would emerge as clear-cut concepts. We may be forgiven for harboring the suspicion that the emergence has not occurred. It therefore remains to recall to mind the various possibilities that would seem to exist for the regulation of embryonic development: that embryogeny is the expression of the inherent capacity of the zygote, that it reflects specific regulation by the environment of the embryo, or that it results from an interaction between the two.

General comment

If the pattern of embryogeny reflects the inherent capacity of the zygote primarily, or even in part, then the zygote must be regarded as a unique cell in the life history of the plant. In the normal course of development, the zygote is, in fact, a unique cell in terms of what it does. However, there are examples where other cells of the embryo sac or the tissues immediately surrounding it can, under certain conditions, develop as embryos. More striking is the development of haploid plants of several species of *Nicotiana* from uninucleate pollen grains in anthers grown in nutrient culture. Stages of development closely resembled those of normal embryogeny (Nitsch and Nitsch, 1969). Furthermore, it has been shown recently that isolated somatic cells can sometimes simulate embryonic development that may be remarkably like that of the normal embryo, or may develop in patterns that are not directly comparable to those of the normal embryo. The important point is that, even though these latter may be aberrant, the product is a normally organized plant, as is often the case when the embryo develops under abnormal conditions. The fact remains that in most cases the zygote does have a greater capacity for embryonic development than other cells of the plant. This suggests that it may be appropriate to examine the zygote, or more properly the egg, for features of organization that are not characteristic of other cells.

Reference has already been made in Chapter Two to the structure and organization of the egg cell cytoplasm in *Gossypium* and the remarkable changes that occur in the zygote within a few hours of fertilization. Even more striking were the fine structural changes reported (Bell, 1963) to occur in cytoplasmic organization during egg maturation and zygote development in the bracken fern *Pteridium*. Mitochondria and plastids were observed to undergo extensive and perhaps complete degeneration, and the remains of these organelles accumulated at the periphery of the egg cytoplasm where they formed a bounding egg membrane. Studies using radioactive tracers indicated that metabolites failed to penetrate this boundary in the egg and young embryo, whereas they accumulated extensively in adjacent cells of the gametophyte. Numerous and complex evaginations developed from the nuclear envelope late in egg maturation and were pinched off in the cytoplasm where they developed as new plastids and mitochondria replacing those that had previously degenerated. In both cotton and bracken the accumulation of large amounts of DNA was observed in the cytoplasm of the egg during maturation. Thus, there is at least limited evidence suggesting that the zygote does have unusual structural properites that might have a bearing upon its special developmental capacities.

In different groups of plants the region in which the embryo develops is highly variable but for each group this environment has recognizable and characteristic structural, chemical, and physical features. The question to be considered is what role do these features play in the regulation of embryogeny. In the seed plants at least, the chemical environment in which the embryo is developing is highly complex; consideration of this fact led to the development of the concept of embryo factors that act as regulators of development. Experimental evidence has now accumulated that numerous factors, both chemical and physical, can substitute for those complex factors that had been held to be highly specific. Furthermore, the embryo can dispense with many of the factors of its environment even though its pattern of development may then be somewhat atypical. The information at hand does not permit a resolution of the enigmatic problems of embryo development, but it does hint at lines of future investigation that might be pursued profitably. The special chemical and physical features of the embryonic environment act perhaps not so much to regulate development as to provide the proper milieu in which the zygote may express its innate genetic capacity. In such a milieu, normally provided in the embryo sac or archegonium but replaceable experimentally, a cell is conditioned in such a way that it is able to respond to stimuli that trigger its further development. The conditions necessary for complete expression of its inherent developmental potentialities must be provided by the embryonic environment—and they normally are. The environment contains factors that operate at later stages of embryonic development and may have a molding effect of sorts, but these later stimuli seem to be of far less consequence than those that trigger the initial development of the embryo. Thus, though the zygote is not a unique cell, at the beginning of embryogeny it does possess special and distinctive characteristics; and though the environment is not unique, it is well adapted to provide all the essentials for the effective expression of the developmental potentialities of the zygote. Thus it is not surprising to find embryos normally developing in embryo sacs and archegonia, but it is no more surprising to find them developing elsewhere.

REFERENCES

Bell, P. R. 1963. The cytochemical and ultrastructural peculiarities of the fern egg. J. Linn. Soc. (Botany). **58**:353–359.

DeMaggio, A. E., and R. H. Wetmore. 1961. Morphogenetic studies on the fern *Todea barbara*. III. Experimental embryology. Am. J. Botany **48**:551–565.

Freeberg, J. 1957. The apogamous development of sporelings of *Lycopodium cernuum* L., *L. complanatum* var. *flabelliforme* Fernald and *L. selago* L. *in vitro*. Phytomorphology **7**:217–229.

Gunning, B. E. S., and J. S. Pate. 1969. Transfer cells: Plant cells with wall ingrowths, specialized in relation to short distance transport of solutes — their occurrence, structure and development. Protoplasma **68**:107–133.

Jensen, W. A., and D. B. Fisher. 1967. Cotton embryogenesis: double fertilization. Phytomorphology **17**:261–269.

Maheshwari, P. 1950. An introduction to the embryology of angiosperms. McGraw-Hill, New York.

Nitsch, J. P., and C. Nitsch. 1969. Haploid plants from pollen grains. Science **163**:85–87.

Norstog, K. 1961. The growth and differentiation of cultured barley embryos. Am. J. Botany **48**:876–884.

Raghavan, V. 1966. Nutrition, growth and morphogenesis of plant embryos. Biol. Rev. **41**:1–58.

Raghavan, V., and J. G. Torrey. 1963. Growth and morphogenesis of globular and older embryos of *Capsella* in culture. Am. J. Botany **50**:540–551.

Rietsema, J., S. Satina, and A. F. Blakeslee. 1953. The effect of sucrose on the growth of *Datura stramonium* embryos *in vitro*. Am. J. Botany **40**:538–545.

Rijven, A. H. G. C. 1952. *In vitro* studies on the embryo of *Capsella bursa-pastoris*. Acta Bot. Neer. **1**:157–200.

Van Overbeek, J., M. E. Conklin, and A. F. Blakeslee. 1942. Cultivation *in vitro* of small *Datura* embryos. Am. J. Botany **29**:472–477.

Ward, M., and R. H. Wetmore. 1954. Experimental control of development in the embryo of the fern *Phlebodium aureum*. Am. J. Botany **41**:428–434.

Ziebur, N. K., R. A. Brink, L. H. Graf, and M. A. Stahmann. 1950. The effect of casein hydrolysate on the growth *in vitro* of immature *Hordeum* embryos. Am. J. Botany **37**:144–148.

FOUR

The Structure

of the Shoot Apex

It is an interesting fact that in plant science the study of development is not equated with embryology. Although the study of embryos has made significant contributions, it is clear that the framework of developmental study in the higher plants has been provided by postembryonic stages. A very important aspect of embryonic differentiation is the establishment of shoot and root apical meristems at approximately opposite poles of the embryonic body. These meristems, whose origins differ somewhat in the various groups of vascular plants, contribute relatively little to the actual development of the embryo; but they are the centers of postembryonic development, and by their continued activity they give rise to the shoot and root systems. The shoot- and root-building activity of these meristems does not represent a mere unfolding of embryonic rudiments; rather it is a true epigenetic accretion of organs and tissues that were not present in the embryo. Thus, all aspects of development—growth and differentiation, histogenesis, and organogenesis—may be investigated in relation to the activity of apical meristems; and the size and accessibility of these formative regions, in comparison to the minute and enclosed embryo, has made them the favored sites for both descriptive and experimental studies of plant development.

In the total development of the primary plant body via its meristems, it is obvious that many processes are taking place simultaneously; and there can be little doubt that these processes are interrelated and that the

37

interaction among them holds many important keys to the understanding of the plant body, its organization, and its integrated development. Thus it is difficult to give serious consideration to the functional organization of the shoot apical meristem without simultaneously examining the process of leaf primordium initiation, and it is almost impossible to consider leaf inception apart from the functional organization of the shoot meristem. It is, therefore, difficult to fit an analysis of these contemporaneous events into the linear sequence of exposition. As this treatment progresses, although particular aspects of development will be dealt with individually, an attempt will be made to compensate for the arbitrary isolation of the parts of an integrated whole by discussing the nature of the interactions among them whenever possible.

Ultimately it is the functional organization of the shoot apex that the student of development seeks to interpret, the mechanism by means of which this formative region remains capable of continued growth while giving rise to organs and tissues that mature. Although it is improbable that structure alone can provide the basis for such an interpretation, it is evident that a functional interpretation must be based upon, or at least be consistent with, the structure of the shoot apex. Accordingly the structural organization will be examined at the outset.

Morphology

The best way in which to acquire an overall appreciation of the positional relationships of the various components of a shoot tip is to remove from it, under a stereoscopic microscope, successively younger leaf primordia in the sequence of their origin. This will ultimately reveal the terminal meristem (Fig. 4.1). Situated at the distal extremity of the axis and surrounded by the youngest leaf primordia, the terminal meristem may have a variety of geometric forms ranging from conical or dome-shaped to flat or even slightly depressed. The diameter of this initiating region at the level of the insertion of the youngest leaf primordium may be as great as 3500 microns in some species of cycads and considerably less than 50 microns in some species of flowering plants. In most species that have been examined, however, apical diameter falls within the 100- to 250-micron range. Furthermore, within the same species, or even within the same plant, there may be considerable variation in dimensions and shape depending upon the age of the plant, the season of the year, and the involvement of the meristem in the formation of a leaf primordium. In most cases, even where the mature shoot is dorsiventral, the symmetry of the terminal meristem is radial; but some dorsiventral shoots, as in some species of ferns and *Selaginella*, are initiated

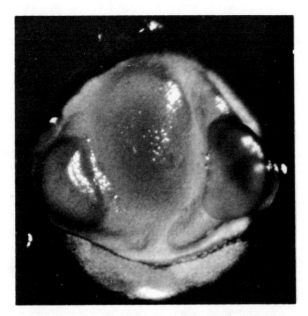

Fig. 4.1   The shoot apex of *Lupinus albus* exposed by dissection. The youngest leaf primordium (P_1) is at the top in this photogrph. P_2 is at the lower left, and P_3 is at the lower right. The first lobes of leaflets are visible on P_3. ×180. (Courtesy D. DesBrisay and J. Waddington.)

by a meristem that is elliptical in shape with its long axis in the plane of dorsiventrality of the shoot.

Histology

Much can be learned from the external examination of apical topography. It is clear, however, that the activity of the terminal meristem involves cellular processes; and it is of paramount importance, therefore, that the cellular organization of the meristem be studied. With the exception of a few investigations in which cellular changes have been examined in living meristems, this study involves the preparation of stained sections from fixed material. The examination of such sections has led to the recognition of an astonishing diversity of structural patterns in terminal meristems. A major difficulty in the interpretation of these patterns is that, whereas function must be understood in terms of the dynamic meristem, the structure can be observed only as it existed at one particular instant in time. This limitation, however, has not interfered with the development of an elaborate typology.

There have been several attempts to produce a workable classification of apical meristem types; perhaps the most often cited is that of Popham (1951) in which seven principal types were designated. On the other hand, one might argue that it is unwise to emphasize pattern types until sufficient functional information is available to indicate which of the structural variations have real significance. In this chapter the approach will be to describe the apices of several species that give some indication of the range of patterns encountered in the vascular plants rather than to define types. In addition to the patterns actually described other forms that illustrate special features will be given brief mention.

Angiosperms—tunica and corpus

It is probable that the shoot apices of flowering plants are more familiar to most biologists than are those of other groups, and it seems appropriate to begin with a representative of this group. The apex of *Lupinus albus* (lupin) has been described by Ball (1949) and has served as the object of a number of experimental investigations. When this apex is examined in median longitudinal section, the region above the youngest leaf primordium and even the youngest two or three leaf primordia themselves present an appearance of cellular homogeneity (Fig. 4.2). The cells are small, nearly isodiametric, thin-walled, and characterized by a high nucleocytoplasmic ratio and by inconspicuous vacuolation. Closer examination reveals that there is a definite stratification of the more superficial regions of the apical mound, which reflects the orientation of planes of cell division in different layers (Fig. 4.3). This is most conspicuous in the two outermost layers of cells of the mound in which divisions above the youngest leaf primordia are ordinarily restricted to the anticlinal plane—that is, at right angles to the surface. In the more deeply seated regions of the apical mound, the planes of cell division are less regularly oriented and in the interior seem to be essentially random. The stratification becomes obscured at the base of the apical mound as foliar primordia are initiated. Furthermore, the cytological homogeneity of the meristem gives way at about the same level to increasing heterogeneity related to the initial differentiation of tissues.

The apical pattern of stratification illustrated by *Lupinus* is characteristic of the angiosperms and is also found in a few other plants. A terminology for the description of such apices, which was devised by Schmidt (1924), is widely used because of its simplicity and general applicability. The name *tunica* has been given to the one or more superficial layers of the apex that, above the level of the youngest leaf primordium, show only anticlinally-oriented cell divisions. To the remainder of the meristem within the tunica the term *corpus* has been applied. The tunica-corpus

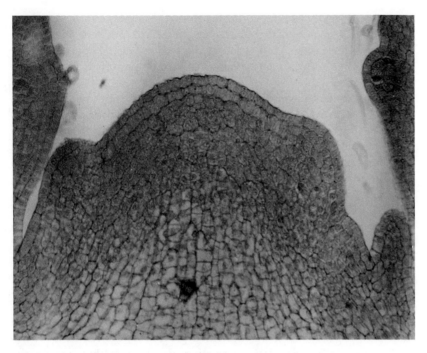

Fig. 4.2   Median longitudinal section of the shoot apex of *Lupinus albus.* ×300.

Fig. 4.3   Diagrammatic illustration of the apex shown in Fig. 4.2. Key: C, corpus; P, leaf primordium; T, tunica. ×300.

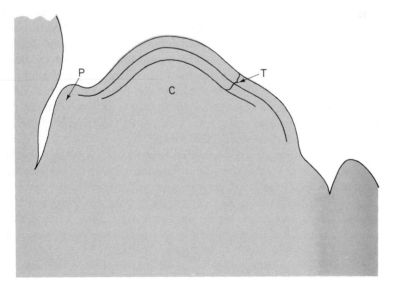

concept has greatly facilitated the description of angiosperm shoot apices; but as more and more of these have been described, problems have arisen concerning the proper application of the terms. This, of course, is to be expected when terminology based upon a static pattern is applied to a dynamic system.

A major difficulty concerns the degree of exclusiveness required in the use of the term *tunica*. Some workers have preferred to follow Schmidt's rather exclusive definition, which restricts the term *tunica* to those layers in which only anticlinal divisions occur. Others have chosen to adopt his actual procedure of admitting an occasional and local periclinal division in a layer designated *tunica*. If the very rigid definition of tunica is applied, some angiosperms, particularly among the grasses, lack even a single tunica layer, although the apical construction is clearly one of stratification. There are also flowering plants in which the entire meristem is stratified, but even the permissive definition allows only two or three of the surface layers to be designated *tunica*. An example of this is seen in *Heracleum* (Majumdar, 1942) in which the meristem consists of eight or nine distinct layers of which only the outermost three constitute tunica.

The number of tunica layers observed in various angiosperms ranges from one to five with the greatest number of species having a two-layered tunica. In *Lupinus*, which was described previously, the tunica is considered to be two-layered. It is interesting also to note that, among the gymnosperms, *Ephedra* and *Gnetum*, as well as several species of conifers, have been found to possess one or more surface layers that correspond precisely to the tunica of the angiosperms, and the tunica-corpus terminology is generally applied in these cases. There are well-documented cases in which the number of tunica layers varies in the same plant, and presumably in the same apex. In *Brassica campestris* (Chakravarti, 1953) the number of tunica layers increased from one in the embryo to three or four in the reproductive shoot. There are indications that the degree of stratification in apices may fluctuate in relation to leaf initiation, reaching a maximum just before the inception of a primordium. In different instances this fluctuation has been interpreted as resulting from changes in the tunica layers or from changes in the division pattern in the outer region of the corpus (Gifford, 1954).

As more and more species have been investigated, it has also become increasingly evident that neither the tunica nor the corpus is cytologically homogeneous. In a number of species a group of cells at the summit of the apex, including both tunica and corpus, have been recognized as being somewhat larger and more highly vacuolate than those around them. These cells, because of their position in the apex, have often been designated *tunica* and *corpus initials*.

The tunica-corpus concept has value in that it provides a framework for descriptive studies of angiosperm shoot apices within which comparisons may be made. The question concerning the limitation of the use of the term *tunica* poses some problems, but it seems that these are best met by a flexible approach rather than by absolute rigidity. Each investigator may then exercise his own judgment in applying the term and has the responsibility of making clear how he has applied it. The biological value of the concept is considerably less evident. Attempts to find correlations between the kind of tunica-corpus organization, ordinarily the number of tunica layers, and phylogenetic position or taxonomic relationships have yielded no consistent results. The significance of tunica and corpus is that they seem to reflect patterns of cell divisions within the meristem and thus to provide information on an important activity of this region. It has been suggested that the different planes of division in superficial and internal regions of the apex may be related to geometric problems of surface and volume growth. This relationship, however, is by no means clear; even if it does pertain, its biological significance is certainly obscure.

Gymnosperms—apical zonation

One of the problems of meristem analysis is the profound differences in structure that are found in apices of different species. This is very well illustrated by comparison of the terminal meristem of *Lupinus* with that of the primitive gymnosperm *Ginkgo biloba* as described by Foster (1938) (Figs. 4.4, 4.5). In *Ginkgo*, although some evidence of stratification may be detected, the structural pattern has a radial organization that centers around a cluster of enlarged, rather highly vacuolated cells that appear to divide infrequently and have been designated the *central mother cells*. The meristem is bounded distally by a *surface layer* in which anticlinal divisions predominate. Periclinal divisions also occur throughout the surface layer, but, together with oblique divisions are considered to be more frequent in the cells at the summit of the apex. These summit cells have been designated the *apical initial group*. Their inner derivatives enlarge and become part of the immediately subjacent zone of central mother cells. Around the zone of central mother cells in a ring-like arrangement is a region of small and densely cytoplasmic cells called the *peripheral subsurface layers*. This region is augmented by periclinal divisions in the overlying surface layer and by division of cells at the margin of the zone of central mother cells. Underlying the zone of central mother cells is a *rib meristem*—that is, a series of rows or files of small, vacuolated cells. The arrangement of these cells suggests an orientation of cell divisions at right angles to the shoot axis. Foster (1938) recognized that between the central mother cells and the cells that

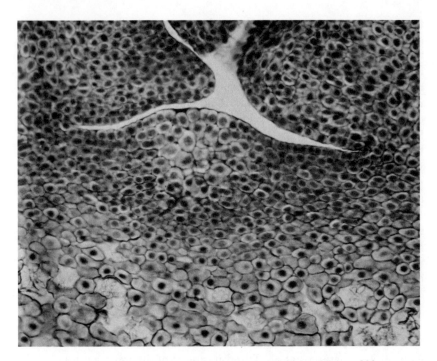

Fig. 4.4   Median longitudinal section of the shoot apex of *Ginkgo biloba.* ×200.

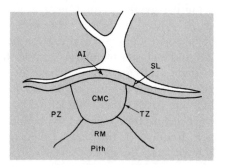

Fig. 4.5   Diagrammatic illustration of the apex shown in Fig. 4.4. Key: AI, apical initial group of cells; CMC, central mother cells; PZ, peripheral sub-surface layers of cells; RM, rib meristem; SL, surface layers of cells; TZ, transition zone. ×200.

surround them laterally and basally there is a *transition zone* in which the cytological characteristics of the central mother cells are gradually replaced by those of the surrounding zones.

Such a pattern, which directs attention to cytological heterogeneity in the meristem, is referred to as a *cytohistological zonation*, and it has been found, with various modifications, to be typical of gymnosperms. The apices of cycads have been interpreted as having this kind of organization, although these remarkable meristems bear little superficial resemblance to *Ginkgo*. In a cycad such as *Microcycas* (Foster, 1943) there appears to be a surface layer over the meristem as in *Ginkgo*, and there is also a large, funnel-shaped group of central mother cells. Between these two there is a region in which short vertical tiers of cells indicate active periclinal division. This region is not present in *Ginkgo*. Foster has designated this region, together with the surface layer, the *initiation zone*. The cells of the central mother cell zone also show a tiered arrangement, which apparently represents the pattern acquired in the initiation zone and retained as cells enlarge and pass into the lower zone. The other regions of the apex are much as in *Ginkgo* except that the peripheral zone is considered to consist of two parts, an outer region derived directly from the initiation zone and an inner region derived from the zone of central mother cells (Fig. 4.6).

A number of descriptions of structural patterns in the shoot meristem in the conifers differ in details. All seem to agree, however, in identifying a distinctive central group of cells beneath the surface layer, surrounded by a peripheral zone (Fig. 4.7). These cells, which are often somewhat enlarged and lightly staining, are sometimes designated *central mother cells* when they are conspicuous and resemble those of *Ginkgo*, and sometimes are called *subapical initials* when they are less distinctive. In any event, it seems clear that a cytohistological zonation is the dominant feature of the shoot apices of most gymnosperms. It will be recalled that some conifers do possess in their shoot apices a degree of stratification that permits the use of tunica-corpus terminology, but this is always superimposed upon a fundamental pattern of zonation of the type just described.

Fig. 4.6   Zonation in the shoot apex of *Microcycas calocoma*. Key: CMC, central mother cells; IPZ, inner peripheral zone; IZ, initiation zone; OPZ, outer peripheral zone; RM, rib meristem; SL, surface layer. (Modified from A. S. Foster. 1943. Amer. J. Bot. **30**: 56.)

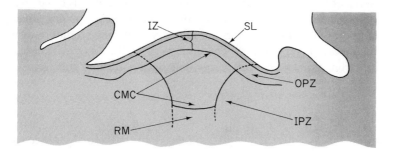

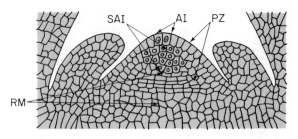

Fig. 4.7   Longitudinal section of the shoot apex of *Pinus palustris* showing the radial zonation pattern in which the central cells are interpreted as apical and subapical initials. Key: AI, apical initials; PZ, peripheral zone; RM, rib meristem; SAI, subapical initials. ×200.

Apical zonation in angiosperms

Following Foster's recognition of cytohistological zonation in *Ginkgo* and the extension of this interpretation to other gymnosperms, it soon was found that apices of certain angiosperms could be more fully described in these terms than according to the tunica-corpus concept alone. For example, Majumdar in 1942 called attention to this kind of pattern in the shoot apex of *Heracleum* and specifically compared it to the structure found by Foster in *Ginkgo*. One of the most striking examples of this type of cytohistological zonation in the apices of angiosperms has been described by Johnson and Tolbert (1960) in several species of *Bombax*. The tunica-corpus terminology can be applied to these apices, the majority having a one-layered tunica. However, the most conspicuous structural feature of the apex is a central mass of cells that resemble closely the central mother cells of *Ginkgo*. This zone extends to the surface and thus includes a portion of the tunica. The cells of this central zone are enlarged and vacuolated and appear to divide rather sluggishly. They are surrounded by a peripheral region composed of small, densely staining cells in a stratified arrangement. Immediately below the central zone is a pith rib meristem. The central zone contributes cells to the zones around it, in which cell division appears to be much more active. Patterns of this type, superimposed upon the tunica-corpus organization, have now been described in many species of flowering plants, although often the cytological characteristics of the central zone are not so distinct as in *Bombax* (Figs. 4.8, 4.9). There seems to be little to distinguish such apices from those of gymnosperms that have a distinct tunica.

Lower vascular plants

When attention is turned to the shoot apices of lower vascular plants a rather different picture emerges from that which has been seen

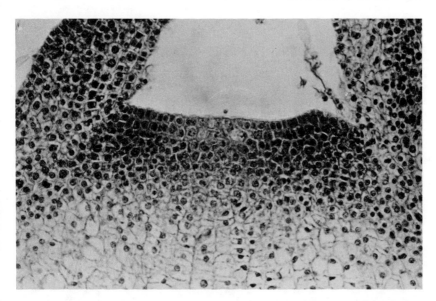

Fig. 4.8   Median longitudinal section of the shoot apex of *Helianthus annuus*. ×235.

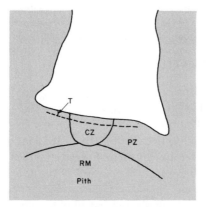

Fig. 4.9   Diagrammatic interpretation of the section shown in Fig. 4.8. Key: CZ, central zone; PZ, peripheral zone; RM, rib meristem; T, tunica. ×235.

in the seed plants. In many of these plants the most prominent feature of apical organization is a superficial layer of axially elongate cells. A fern such as *Osmunda cinnamomea* (cinnamon fern) will serve to illustrate this type of pattern (Steeves, 1963) (Fig. 4.10). In *Osmunda* the superficial cells are rectangular in shape and are lightly stained by ordinary histological dyes. The center of the surface layer is occupied by a somewhat larger cell that is pyramidal in shape, and which has been called the

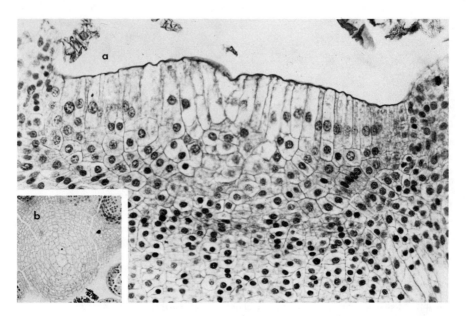

Fig. 4.10 Shoot apex of *Osmunda cinnamomea*. (a) Median longitudinal section in which the apical cell is visible in the center of the surface layer of enlarged cells. Its relationship to the surface cells may also be seen in the cross section of the apex. (b) Cross section of the apex. Immediately below the surface layer of the apex in longitudinal section, or outside the surface layer in cross section, the start of histodifferentiation can be seen. (a) ×82, (b) ×17. (T. A. Steeves. 1963. J. Indian Bot. Soc. **42A**: 225.)

apical cell. This cell divides only anticlinally and its derivatives augment the surface layer. The other cells of the surface layer divide both anti-clinally and periclinally, but the periclinal divisions are unequal, so that small inner derivatives are cut off and the distinctive shape of the surface cells is retained. However, at the margins of the apical mound these cells become segmented and lose their distinctive appearance. The apical cell in *Osmunda* is not very distinct morphologically; and during periods of active growth it can be distinguished only with difficulty because its segmentation pattern is irregular. There is some doubt as to whether a single cell functions continuously as the apical initial. Immediately beneath the surface layer, histological differences become apparent among the cells in different regions, and these have been interpreted as repre-senting the initial stages of histodifferentiation.

   This type of apical organization seems to be general among the lower vascular plants. In all of these a superficial layer of relatively enlarged cells functions as the initiating region, and immediately below this the initial stages of tissue differentiation may be detected. In some cases the apical cell is enlarged and very conspicuous as in *Equisetum* (Fig. 4.11) and many ferns. In other cases, a group of initial cells replaces the single apical cell, and considerable variation may be found in the same plant, or perhaps even in the same apex at different times. *Osmunda*, just described, is of this second type. Shoot apices of *Lycopodium* also can be included in this group even though the cells of the superficial layer are not so conspicuous in size and vacuolation as in other lower vascular plants.

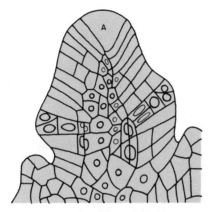

Fig. 4.11   Diagrammatic illustration of a median longitudinal section of the shoot apex of *Equisetum arvense*. Key: A, apical cell. ×275. (S. J. Golub and R. H. Wetmore. 1948. Amer. J. Bot. **35** : 755.)

General comment

What seems to emerge rather forcefully from this comparative survey of shoot apical organization in various groups of vascular plants is the fact that structural patterns are diverse, perhaps even more diverse than the mature shoots to which they give rise. This circumstance makes it very difficult to interpret the structural patterns in functional terms, and indeed, one begins to wonder if they have functional significance at all. It is not, however, legitimate to separate function from structure; and any interpretation offered at the functional level must, at least, be consistent with the morphology and histology of the apical region. Before proceeding with a discussion of analytical and experimental investigations that have explored the workings of the shoot apex, it may be appropriate to conclude this chapter with a brief account of one rather widely applied interpretation that, although histological in nature, has definite functional implications.

Throughout the history of the study of shoot apices, most investigators have identified, or at least designated, certain cells at the summit of the apex as *apical initials*, and this term has appeared several times in descriptions given earlier in this chapter. In many lower vascular plants a distinctive apical cell may be recognized, which by its regular segmentation appears to function as an initial cell for the entire shoot, while retaining its individual identity. In this sense it functions as the ultimate, although of course not the immediate or direct, source of all of the cells of the shoot. In other lower plants, several such initial cells seem to fulfill this role. Undoubtedly the conspicuous presence of one or more initials in the apices of lower plants was an important factor in the widespread acceptance of the existence of apical initials even where they are not cytologically distinctive, as in the seed plants. The concept of a special group of cells, presumably in a particular physiological state, that imparts to the shoot meristem its property of continued meristematic activity is an attractive one, and one that has been inherent in most descriptive studies of the shoot apex.

The presence of apical initials usually has been associated, if not directly, at least by implication, with another feature of the organization of the apex, which is often designated the *promeristem* (Sussex and Steeves, 1967). The promeristem is defined as including the apical initials and their most recent derivatives, which have not as yet undergone any of the changes associated with tissue differentiation. The recent derivatives of the initials are presumed to remain in more or less the same physiological state as the initials, but because of their position they, unlike the initials, are not totipotent. Thus, although it has important functional implications, the promeristem is a histological concept based upon the absence of true histodifferentiation. It has definite structural features that vary in different plants, but these features, unlike those of immediately subjacent   regions, are not related to the differentiation of the tissues of the shoot. A major problem in the application of this interpretation, however, is the extreme difficulty in recognizing the initial stage of tissue differentiation and consequently the boundaries of the promeristem, if indeed these boundaries can be thought of as being sharply defined. Rarely in studies of apical zonation has there been any attempt to designate which of the features observed represent the beginnings of the cytological changes associated with tissue differentiation. In several ferns it has been suggested that tissue differentiation can be detected immediately beneath the surface layer of enlarged cells, and the surface layer itself has been designated as the promeristem.

The concept of a promeristem composed of cells that have not undergone tissue differentiation, surrounding one or more initial cells that confer upon the apex its ability for continued growth, seems to be in accord with the structural patterns already described; at least it does not conflict with them. It is, in fact, no more than a working hypothesis based upon observed structure; and it is in no sense a true functional interpretation explaining how the shoot apex performs its all-important role in shoot growth. Moreover, it has been challenged by other interpretations, and particularly in recent years by a growing body of analytical and experimental information. It remains now to examine some of this evidence against the background of the traditional interpretation.

REFERENCES

Ball, E. 1949. The shoot apex and normal plant of *Lupinus albus*, bases for experimental morphology. Am. J. Botany **36**:440–454.

Chakravarti, S. C. 1953. Organization of shoot apex during the ontogeny of *Brassica campestris* L. Nature **171**:223–224.

Foster, A. S. 1938. Structure and growth of the shoot apex in *Ginkgo biloba*. Bull. Torrey Botan. Club **65**:531–556.

———. 1943. Zonal structure and growth of the shoot apex in *Microcycas calocoma* (Miq.) A.DC. Am. J. Botany **30**:56–73.

Gifford, E. M., Jr. 1954. The shoot apex in angiosperms. Botan. Rev. **20**:477–529.

Johnson, M. A., and R. J. Tolbert. 1960. The shoot apex in *Bombax*. Bull. Torrey Botan. Club **87**:173–186.

Majumdar, G. P. 1942. The organization of the shoot in *Heracleum* in the light of development. Ann. Botany (London) (NS) **6**:49–82.

Popham, R. A. 1951. Principal types of vegetative shoot apex organization in vascular plants. Ohio J. Sci. **51**:249–270.

Schmidt, A. 1924. Histologische Studien an phanerogamen Vegetationspunkten. Botan. Arch. **8**:345–404.

Steeves, T. A. 1963. Morphogenetic studies of *Osmunda cinnamomea* L. The shoot apex. J. Indian Botan. Soc. **42A**:225–236.

Sussex, I. M., and T. A. Steeves. 1967. Apical initials and the concept of promeristem. Phytomorphology **17**:387–391.

FIVE

Analytical Studies of the Shoot Apex

If the study of structural patterns in the shoot apices of vascular plants does not lead to an understanding of this region in functional terms, other methods of investigation must be employed to attain such an understanding. A considerable body of research has attempted to analyze more precisely the activities of the shoot tip and its component parts and has produced some new information and brought some of the more challenging unsolved problems into sharper focus. This chapter will discuss some of the significant contributions that analytical studies have made.

Frequency and distribution of mitosis in the shoot apex

One essential function of the shoot meristem is that of producing the cells of which the stem and leaves are composed; and the distribution of cell divisions within the meristem has long been recognized as an important problem. It is important to know whether cell division is more or less uniformly distributed throughout the terminal region of the shoot or whether it tends to be localized in certain cell-producing regions, with the remaining regions of the meristem having other functions. One might also ask to what extent the organizational patterns that have been noted are related to the distribution of cell divisions in the apex.

The presence of a cluster of enlarged and somewhat vacuolated cells in the center of the meristem in the shoot apices of a number of vascular plants has suggested to some workers in the past that this region may be characterized by a rate of cell division somewhat slower than that in surrounding regions (Philipson, 1954). Despite the presumed low rate of division in this region, however, it includes the cells, variously designated as tunica and corpus initials or apical and subapical initials, which have been regarded as the ultimate source of all the cells of the primary body of the shoot. Early observations on this region did not usually include any quantitative estimates of rates of cell division in various zones of the apex, but were based largely upon visual inspection. It came, therefore, as something of a shock to students of plant development when Buvat and his coworkers in France began to report the results of actual counts of mitotic figures in various regions of shoot apices of a number of species, for these indicated the presence at the very summit of the apex of a region devoid of mitotic activity (Buvat, 1952; Lance, 1952). This concept of apical organization elaborated by the French school had its origins in a theory of phyllotaxy suggested by Plantefol in which it was asserted that leaf initiation is regulated by older regions of the shoot through stimuli propagated along *foliar helices* to a peripheral meristematic ring in the apex in which the leaf primordia are initiated. The method used by Buvat and his associates was the construction of composite diagrams in which the positions of mitotic figures from a number of microtome sections were superimposed within the outline of a single drawing. In the case of transverse sections, serial sections of the same apex were superimposed, and in the study of longitudinal sections, near median sections of a number of apices were superimposed. By this means it was possible to overcome the difficulty posed by the fact that most individual sections show very few mitotic figures; and enough locations could be accumulated in a single diagram to reveal a pattern of distribution. Such diagrams revealed to the French workers a terminal zone of the apex in which mitoses were absent, or extremely rare. On the basis of these results, and supporting cytological evidence, a revolutionary theory of apical organization was proposed (Fig. 5.1).

According to this theory, the summit of the apex in a vegetative shoot is occupied by a distinctive zone, called the *méristème d'attente*, which plays no histogenetic or organogenetic role in the development of the leafy shoot—this being reflected in the absence or near absence of mitotic activity in the zone. The méristème d'attente is surrounded laterally by the *anneau initial* and is subtended by the *méristème médullaire*, both of which are characterized by the occurrence of frequent mitotic figures. To the

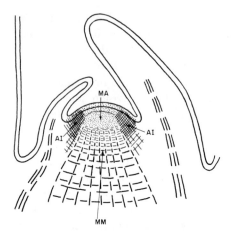

Fig. 5.1 Diagram of the shoot apex of *Cheiranthus cheiri* interpreted according to the *méristème d'attente* concept. Key: AI, *anneau initial*; MA, *méristème d'attente*; MM, *méristème medullaire.* (R. Buvat. 1955. Ann. Biol. **31** : 595.)

anneau initial is ascribed the major role in the development of the shoot in that it gives rise to leaf primordia and the associated stem tissues. The méristème médullaire gives rise to the cells of the pith. This organization, which prevails throughout vegetative growth of the shoot, changes drastically with the onset of floral or inflorescence development. Abundant mitoses may be detected in the méristème d'attente at this time; and the previously inactive region gives rise to most or all of the reproductive structures. According to the theory, this is the function of the méristème d'attente for which it "waits" throughout vegetative development. The different regions of the apex may be rather indistinct in early growth phases of the plant. It is reported that in some species the regions of the apex are not clearly recognizable at the time of seed germination but become progressively more distinct as vegetative development progresses.

The hypothesis was based originally upon the examination of apices of several species of dicotyledons, notably *Cheiranthus cheiri*, *Lupinus albus* (lupin), *Myosurus minimus*, and *Vicia faba* (broad bean); but it soon was extended to monocotyledons such as *Triticum vulgare* (wheat) and *Luzula pedmontana* and to several gymnosperms including *Picea excelsa* and *Ginkgo biloba* (Camefort, 1956). The distinctive meristems of the lower vascular plants have received far less attention, but in *Equisetum arvense* the theory is held to be applicable. It has now become apparent to the proponents of the méristème d'attente theory that there are many cases, particularly among the gymnosperms, in which the inactive region never gives rise to reproductive structures, as had been visualized in the dicotyledons originally examined. Presumably in such cases the inactive cells never become activated; and one is left with the apparent anomaly of a group of cells in the center of the meristem that have no function at all in the life of the plant.

The interpretation of shoot apical organization proposed by the French school gained very limited acceptance elsewhere. Initially objections stemmed from the fact that many workers had observed mitotic figures or evidence of recent mitoses in the summit cells of a variety of shoot apices (Esau et al., 1954; Popham, 1958), which raised questions about the inactivity of the méristème d'attente. On the other hand, because of the short duration of the actual mitosis in the total cycle of cell reproduction, it is difficult to observe mitoses in a small cell population unless the mitotic index is high; and observations of shoot tip sections often fail to reveal mitoses at the summit of the apex. This difficulty may be overcome by the use of radioactive labeled compounds such as tritiated thymidine or adenine, which are incorporated into the nuclei during the synthetic phase of the cell cycle. Because the cells are exposed to the labeled compound for a period of hours, the investigator is not dependent upon the fortuitous occurrence of a mitosis at the time of fixation of material. Apices treated in this fashion have in a number of studies revealed labeled nuclei in cells at the summit of the apex (Clowes, 1961).

The demonstration of some mitotic activity at the summit of the apex does not in itself invalidate the theory, as the French workers themselves have reported the same observations and have taken it into account in formulating the theory. In its original statement, the méristème d'attente theory pictured the shoot apex as consisting of a dormant cap or dome of cells continually carried forward by proliferation in the anneau initial and the méristème médullaire. As the theory has been developed, the occurrence of a limited amount of mitotic activity in the méristème d'attente has been recognized at least in some cases, and along with this some interaction among the zones of the apex. The inception of a leaf primordium in a particular region of the anneau initial reduces the extent of this zone in that part of the apex, and this is followed by the restoration of the anneau initial to its original dimensions. This process, which has been referred to as *regeneration*, involves increased mitotic activity in a localized region of the anneau, and this activity may encroach upon the adjacent region of the méristème d'attente. Thus, in some species, mitotic figures may be observed in the méristème d'attente from time to time in localized regions. This is especially significant in small apices. For example, in the apex of *Luzula pedmontana*, the encroachment of mitotic activity causes an apparent shifting in the position of the méristème d'attente, which is in fact a shifting of mitotic activity rather than a movement of the zone itself (Catesson, 1953). In *Scabiosa ukranika* the méristème médullaire also encroaches upon the méristème d'attente with the result that there is no permanently inactive region. A similar explanation was applied to the apex of *Equisetum* in which an occasional mitotic figure may be seen in the apical cell itself.

In essence, the theory visualizes the anneau initial as the generative center of the vegetative shoot, with a supplementary contribution by the méristème médullaire. The summit of the meristem, which had previously been considered to be the ultimate source of all the cells of the shoot, is here assigned a purely passive role. In some cases it is considered to remain totally inactive whereas in others it may manifest a limited amount of mitotic activity, but only under the influence of proliferative stimuli received from more basal regions.

Implicit in many of the criticisms of the work of the French school has been the idea that, although mitoses do occur in the summit cells of the apex, their frequency is low. This idea, in fact, preceded the méristème d'attente theory and was based, as has been pointed out, upon the recognition of cytohistological zonation in many apices. There are several studies, apart from those of the French school, reporting counts of mitotic figures in the various zones of the apex that indicate a lower mitotic frequency in the central region than in the peripheral and basal zones. Such a result has been obtained in plants as diverse as *Elodea* (Savelkoul, 1957) and *Coleus* (Jacobs and Morrow, 1961). Similarly, apices of *Arabidopsis* supplied with tritiated thymidine incorporated significantly less of this DNA precursor in the central zone than in the peripheral areas of the apex (Brown, et al., 1964). Gifford and his colleagues (1963) were led to a similar conclusion in a study of several other species. This is almost exactly the result obtained by supporters of the méristème d'attente theory in recent studies (Saint-Côme, 1966; Nougarède et al., 1964). Although they acknowledge that some labeled nuclei are found in the méristème d'attente, particularly during the regenerative phase, they always find significantly fewer in this region than in the anneau initial in the vegetative apex (Table 5.1). In a recent review Nougarède (1967) has concluded that a brief exposure to labeled precursors (approximately six hours) reveals the méristème d'attente very clearly, but that longer exposure such as was used in the studies referred to earlier obscures the pattern as cells of the central zone gradually become labeled. Interestingly enough, some of the strongest autoradiographic evidence in support of the méristème d'attente has come from a study of shoot apices of *Helianthus annuus* (sunflower) (Steeves et al., 1969), which was not done by the French school. Apices of this species were excised at various stages of development, placed in sterile culture, and supplied with tritiated thymidine via the culture medium. No detectable incorporation was noted in the central zone throughout the period of vegetative development even when the period of exposure was increased to 48 hours

TABLE 5.1

MITOTIC FREQUENCY AND THE INCORPORATION OF THYMIDINE-H^3 IN THE ADULT, VEGETATIVE SHOOT APEX OF COLEUS BLUMEI BENTH. EACH FIGURE REPRESENTS THE SUMMATION OF RESULTS FROM TEN APICES. BASED ON DATA FROM SAINT-CÔME, 1966.

Region of apex	*Nonregenerative phase*			*Regenerative phase*		
	MITOTIC FREQUENCY					
	No. of mitoses	Total no. of nuclei	% of nuclei in mitosis	No. of mitoses	Total no. of nuclei	% of nulei in mitosis
Apical zone (*méristème d'attente*)	2.0	376	0.5	6.0	161	3.7
Peripheral zone (*anneau initial*)	7.0	225	3.1	45.0	463	9.8
	INCORPORATION OF THYMIDINE-H^3*					
	No. of labeled nuclei	Total no. of nuclei	% of nuclei labeled	No. of labeled nuclei	Total no. of nuclei	% of nuclei labeled
Apical zone (*méristème d'attente*)	5.0	295	1.6	10.0	252	4.0
Peripheral zone (*anneau initial*)	10.0	180	5.7	52.0	553	9.4

*Applied directly to the apex at a concentration of 30 microcuries per milliliter.

(Fig. 5.2). Though the authors of this report were not willing to exclude the possibility of some DNA synthesis and mitosis in the central zone cells during vegetative growth, they were forced to conclude that, at the very least, their rate must be extremely slow.

One fact that must be kept in mind in evaluating all of this work is that it is unrealistic to base conclusions about mitotic frequency in the several zones of the apex upon counts of mitotic figures, or even of labeled nuclei, without relating these counts to the number of cells present in the region counted. Because cells in the center of the meristem are known to be larger in many cases than those of more peripheral regions, it might be expected that the number of cells encountered in division would be small, and plots of mitotic frequency in the meristem could, in fact, be nothing more than plots of nuclear frequency. Some of the early quantita-

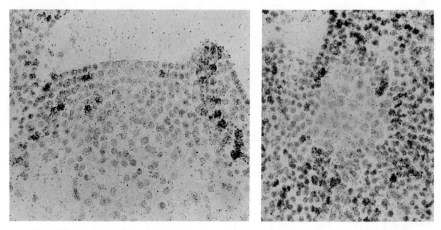

Fig. 5.2 Median longitudinal (left) and transverse (right) sections of the shoot apex of *Helianthus annuus* after feeding with tritiated thymidine, and autoradiography of sections. Nuclei that had synthesized DNA during the feeding period may be identified by the accumulation of black silver grains over them. Note the absence of silver grains over nuclei in the central zone. The transverse section was cut slightly below the surface of the apex. ×260. (T. A. Steeves et al. 1969. Can. J. Bot. **47**: 1367.)

tive studies on cell division frequency in shoot apices were justifiably criticized for failing to consider this important limitation. More recent investigations, however, have expressed their results in terms of percentages; and this factor no longer can be held to invalidate the conclusions.

Thus it appears that there is growing support for the concept of a central region in the shoot apex, at least of seed plants, in which DNA synthesis and mitosis are greatly restricted during vegetative growth. By contrast, at least in angiosperms, mitosis becomes very active in this region with the onset of reproductive development. In this sense the claims of the French school have gained acceptance; but in fairness it must be pointed out that it was not in this sense that these claims were criticized in the first place. A very fundamental objection to the méristème d'attente concept still remains, and it concerns the significance of the divisions that do occur in this region of the apex. The French workers have treated these as passive byproducts of developmental processes in the anneau initial, thought by them to be the true generating center of the shoot. Critics (Wardlaw, 1957), however, have pointed out that a small group of cells at the summit of an apex would need to divide only very infrequently in order to provide constant replenishment for the organogenetic and histogenetic regions that subtend them and that actually produce the organs and tissues of the shoot. Thus the true initials of the shoot need not divide very frequently in order to serve as the ultimate source of all the cells of the shoot. In a study of the shoot apex of *Ephedra altissima* Paolillo and Gifford (1961) have concluded that if the subapical initials divide periclinally only once during each plastochron, the time period between initiation of two successive leaf primordia, this would be sufficient to account for the increase in height of the apex that occurs during the same period because of the subsequent divisions and expansion of the derivative cells.

Although rather substantial agreement seems to have been reached regarding the histological evidence for the occurrence of significantly different cell division frequencies in the several zones of the shoot apex, a somewhat different and very promising approach to the question of cell proliferation in shoot apices, pioneered by Newman (1956) and by Ball (1960), has led to very different conclusions. This new approach is the direct observation of cell division patterns in living shoot apices. In these studies sequential observations of cells in the surface layer of the meristem of several species of angiosperms were made and recorded by drawings or by photography. Changing cellular configurations and the appearance of new walls at or near the summit of the apex provided convincing evidence that cell division does occur in this region of the apex. Ball's photographic records indicate that the frequency of division is high in the summit cells of the apex. For example, in one apex of *Asparagus* he found that in approximately 17.5 hours, nine cells originally placed under observation all divided once and three derivatives also divided, producing a final total of 21 cells (Fig. 5.3). Other apices showed less rapid division, but in no case was there any evidence for an axial group of nondividing or sluggishly dividing cells.

On the other hand, there are several limitations to the methods used in

Fig. 5.3   Record of cellular changes in the surface layer of the shoot apex of *Asparagus officinalis.* The drawings are tracings of time lapse photographs. Nine cells that could be identified at zero time are numbered, and these numbers are retained for the derivatives resulting from their divisions. Thus in the first picture there is one cell designated 5. 10.1 hours later it had given rise to two cells, each labeled 5, and 7.4 hours later one of these had divided again so there are then three cells labeled 5. ×300. (Drawn from E. Ball. 1960. Phytomorphology **10**: 377.)

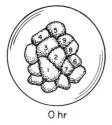
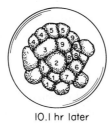
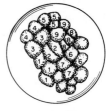

O hr   l0.l hr later   7.4 hr later

this study that prevent a completely uncritical acceptance of the results. The only cells observed were those at the summit of the apex, and no comparison was actually made with cells in more lateral positions. Thus relative rates of division in the meristem remain unknown. Moreover, it would seem that it is very difficult to judge exactly the location of the summit cells in a domed apex, especially because the summit may shift during the plastochron. Finally, in order to make the observations, the shoot tips were excised and grown in sterile culture and many leaf primordia surrounding the meristem were removed. It is difficult to evaluate the possible influence of these rather drastic treatments, although it may be recalled that somewhat comparable operations in the case of *Helianthus* shoot apices did not cause incorporation of tritiated thymidine in nuclei of the central zone. Nonetheless, it is abundantly clear that these observations of living meristems must be given careful consideration in any attempt to formulate generalizations about the distribution of cell divisions in shoot apices.

The idea that the cells at the summit of the apex, even though they may divide occasionally, play no important role in organogenesis or histogenesis, as is maintained by the French school, has been challenged further on the basis of another type of evidence, that provided by the study of polyploid chimeras. In a polyploid chimera, tissues of different ploidy are mixed in various ways. In a periclinal chimera, parallel layers of cells with different ploidy levels are found, as for example in a shoot that is entirely diploid except for a tetraploid epidermis. In a sectorial chimera, tissues of a particular ploidy level occupy a sector of the organ, and in a mericlinal chimera a periclinal chimera occupies only a sector of the organ. Chimeras may arise spontaneously or they may be induced, in the case of polyploid chimeras, by treatment of the shoot apex or the seed or seedling with a dilute solution of colchicine. Polyploid chimeral plants can be recognized after such treatment by particular histological or cytological features in the affected tissues, as for example enlarged stomatal guard cells in the case of a polyploid epidermis. When chimeras are relatively persistent, it is inferred that they result from correspondingly stable changes in the apical meristem; and in fact, such changes can often be seen in the meristem in the case of periclinal chimeras. In *Datura*, Satina, Blakeslee, and Avery (1940), by studying periclinal polyploid chimeras, were able to identify three layers in the meristem, each of which responded independently to treatment with colchicine (Fig. 5.4). Each of the meristem layers could be related to a layer or layers of tissue in the resulting mature plant body by a comparison of ploidy levels. These tunica and corpus layers are most distinctive at the summit of the apex, and their close relationship to tissues of the mature body of the shoot has been taken as evidence that the summit cells in the apex have a true

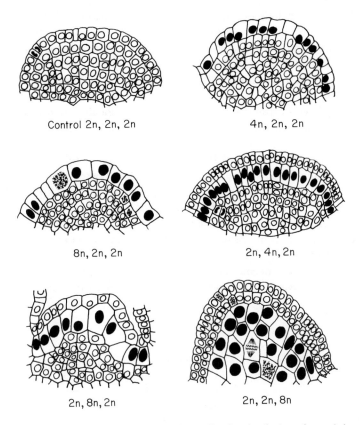

Control 2n, 2n, 2n   4n, 2n, 2n

8n, 2n, 2n   2n, 4n, 2n

2n, 8n, 2n   2n, 2n, 8n

Fig. 5.4 Periclinal polyploid chimeras seen in median longitudinal sections of the shoot apex. The ploidy level of each of the three apical layers is shown for each apex. Diploid nuclei are shown as open circles and polyploid nuclei as filled circles. (Redrawn from S. Satina et al. 1940. Amer. J. Bot. **27**: 895.)

initiating role. Buvat (1955), however, arguing for the passive role of the summit cells, has insisted that, in the absence of precise cytological information as to the origin of the meristem layers, it is possible that the polyploid cells arise in the anneau initial and that the pattern in the summit cells resulted from processes involved in regeneration of the anneau.

A slightly different but related problem has been approached by the use of induced mericlinal chimeras in cranberry. Dermen (1945) found that in epidermal mericlinal chimeras, the altered segment occupied approximately one third, one half, or two thirds of the stem. This suggested that in the apex there must be two or three apical initial cells. The development of polyploidy in one of these would thus be expressed in one third or one half of the shoot only, the remainder being normal. Such a situation clearly would require the activity and the importance of the summit cells of the apex. The chimeral patterns varied in permanence and often disappeared after a period of growth, indicating that if such apical initials are present they are not to be regarded as permanent initials but could be displaced by shifting configurations in the apex. Moreover, because there is little positive cytological evidence as to the mode of origin of such chimeras, it is difficult to be certain of their precise significance.

Experiments by Soma and Ball (1964) also bear upon the question of the permanence of initial cells in the meristem. These authors, working with *Lupinus*, applied small carbon particles to the summit of the apex or, alternatively, made small needle punctures in the same region and they observed the subsequent displacement of these markers. Records of the subsequent positions of these markers indicated that many of them were displaced from the apical summit, and this has been interpreted to mean that there are no permanent apical initials occupying that position (Table 5.2). However, it is worth noting that after 21 days more than 20

TABLE 5.2

THE FATE OF THE CARBON SPOTS THAT HAD BEEN PUT ON THE EXACT CENTER OF SHOOT APICES OF LUPINUS ALBUS AS ASCERTAINED AFTER VARIOUS NUMBERS OF DAYS OF SUBSEQUENT GROWTH (FROM SOMA AND BALL, 1964).

	Location of carbon spot (numbers indicate %)					
Days after treatment	*Shifted to one side of flanks*	*Divided during the movement*	*Scattered around apical center*	*No visible change*	*Missing*	*No. plants studied*
21	47.3	10.9	18.2	23.6	0	55
28	50.0	12.5	12.5	12.5	12.5	8
42	60.0	6.7	10.0	6.7	16.6	30

percent of the apices marked with carbon particles showed the marker still located at the center, and after 42 days more than 5 percent still showed no displacement. In view of the difficulty of ascertaining the exact center of the meristem and the fact that a significant percentage of the apices showed a very slow displacement, these results must be taken as highly suggestive, but not conclusive indications of the absence of permanent or long-term apical initials.

In spite of the considerable number of investigations devoted to the question of the distribution of cell divisions in the shoot meristem, and particularly to the localization of regions of relative activity and inactivity and the significance of such regions if they exist, it is quite evident that these problems have not been solved. It would seem that the methods thus far applied to these problems ought to be adequate for their resolution, and it may be that by applying several of the methods conjointly to the same species, apparent discrepancies can be eliminated.

Cytohistological studies

It now should be recalled that there are other indications of localized physiological activity within the shoot apex provided by the various zonation patterns noted in the previous chapter. In addition to differences in cell size and planes of cell division, these patterns have been shown to have a basis in the kinds of cytological features that suggest differences in metabolic activity. Whether or not these differences are related to or correlated with differences in mitotic activity, as some workers maintain, remains to be established. Workers of the French school, in support of the concept of the méristème d'attente, have carried out detailed cytological studies on a variety of angiosperms and gymnosperms in which they have shown important localized differences within the meristem (Buvat, 1955; Nougarède, 1967). The summit cells of the apex are characterized by large vacuoles, filamentous mitochondria, small nucleoli, and the presence of differentiated plastids, whereas those of the anneau initial possess small vacuoles, granular mitochondria, large nucleoli, and undifferentiated plastids. These cytological features have been associated with the differences in mitotic activity of the cells in the two regions. Within the anneau initial itself, there are regional differences. Those cells in the process of initiating a leaf show the anneau characteristics to the fullest extent, whereas those in less active regions are intermediate between the anneau initial condition and that of the méristème d'attente. Cytochemical tests carried out on shoot apices have shown localized differences in the cellular content of RNA which is low in the méristème d'attente and high in the more peripheral regions of the meristem. Incorporation studies using radioactive precursors have indicated that the synthesis of RNA is low in the axial cells of the meristem as compared to the surrounding regions. Studies with the electron microscope have generally verified these observations at the optical level and in addition have substantiated the conclusions about RNA concentration by revealing a higher density of ribosomes in the peripheral zone than in the axial cells of the apex (Nougarède, 1967). The very limited occurrence of DNA synthesis in the nuclei of the central zone has already been noted; and these nuclei, which are relatively large, give a conspicuously faint response to the Feulgen reaction, which is specific for DNA (Fig. 5.5). However, under spectrophotometric analysis, these nuclei do not appear to have less DNA than other diploid cells (Steeves et al., 1969). The pale staining reaction appears, therefore, to result from DNA dilution in these enlarged nuclei.

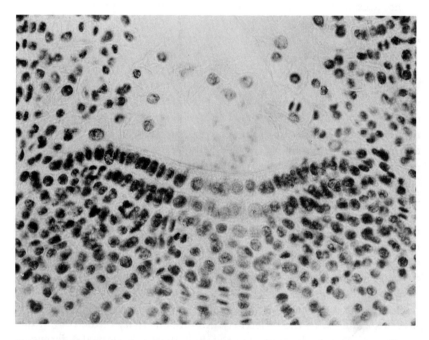

Fig. 5.5   Median longitudinal section of a shoot apex of *Helianthus annuus* treated by the Feulgen reaction which stains only DNA. The large nuclei in the central zone are less intensely stained than are those in the peripheral zone. ×420. (T. A. Steeves et al. 1969. Can. J. Bot. **47**: 1367.)

Gifford and Tepper (1962a, b) have made observations on the shoot apex of *Chenopodium album* that are somewhat different from those of the French school. In the seedling they found a zonation based upon the kinds of cytological characteristics emphasized by Buvat and his associates, but this gave way to a much greater uniformity in the adult apex. Here only the axial tunica cells showed larger vacuoles, more differentiated plastids, and more filamentous mitochondria than were present in the surrounding cells, but the nucleoli were larger and a difference in RNA content was not evident. However, the nuclei of the axial tunica cells did appear to be lower in DNA and histone content than did those of more peripheral cells.

All of these cytological features are typical of the vegetative apex only; and the initiation of flowering, often after a very brief period of induction in the case of photoperiodically sensitive species, results in the development of uniformity throughout the terminal meristem as cells of the central zone acquire the characteristics previously found only in the peripheral zone. This, of course, is associated with the acceleration of mitotic activity in this region. Some recent work on photoperiodically sensitive plants has revealed an interesting aspect of these changes, that may go a long way toward explaining some of the differences in staining patterns reported by earlier workers. It now appears that if such a plant is held in a light regime that is unfavorable for the initiation of flowering, the shoot apex may nonetheless undergo some of the cytological changes associated with flowering while still remaining vegetative in function (Nougarède et al., 1965). Thus, some of the zonation, or lack of it, that has been attributed to vegetative apices may in fact be applicable to this transitional stage, and great caution must be exercised in the interpretation of such results.

Metabolic differences in various regions of a shoot apex ought to be reflected in patterns of enzyme distribution; and it is quite possible that some of the earlier described histological zonation in meristems might be related to such patterns. There has been very little work dealing with this aspect of meristem organization, but a few studies may be reported. Wetmore, Gifford, and Green (1959) found, in the apices of some species, that histochemical tests for oxidases and dehydrogenases gave indications of greater amounts of both in the mitochondria of the centrally located cells of the meristem than in peripheral and subjacent tissues. More recently Vanden Born (1963) has carried out extensive observations on sections of unfixed apices of *Picea glauca* in which attempts were made to localize a number of enzymes in the meristem and subjacent tissues. The most interesting pattern to appear was that of peroxidase, which was abundant in the peripheral region of the meristem and the immediately subjacent tissues as well as in differentiating vascular tissue and in other regions in which mitotic activity is believed to be high. It was significantly absent from cells at the summit of the apex, and thus from the zone that corresponds to the méristème d'attente. It is difficult to interpret results of this type, but they do indicate that a more physiological approach to the problem of meristem organization is possible.

To date, cytochemical work on shoot apices has obviously been too fragmentary to permit any general conclusions to be drawn concerning localization of metabolic activities in various regions of the meristem. Despite the discrepancies between individual investigations, the methods thus far employed clearly show localizations within the meristem at least under certain conditions, and to this extent they hold promise for the future. In view of what has been reviewed earlier concerning the distribution of cell divisions in the meristem, it is interesting that much of the cytochemical work does seem to point to a central group of cells that are metabolically distinct. The cytochemical differences that have been noted in this region do not necessarily relate to differences in mitotic activity and may be quite independent of such differences if these actually exist.

Biochemical analysis

Another approach to the understanding of physiological characteristics of the shoot apex has made use of more conventional analytical methods. Steward et al. (1954, 1955) carried out a chromatographic analysis of the nitrogenous components of shoot apices of several species. The apex of the fern *Adiantum*, as compared with the apex of the angiosperm *Lupinus*, was found to be relatively poor in soluble amino acids. In *Lupinus* marked differences were noted between the meristem and mature tissues derived from it. The meristem was characterized by having a higher content of basic amino acids in both soluble and protein fractions than the more mature regions have. This analysis had as its objective, not the localization of metabolic activities within the meristem, but rather the overall characterization of the meristem in relation to its mature derivatives. Nonetheless, if the methods could be further refined, they might provide an interesting approach to the physiological organization of the meristem.

An analysis of a somewhat different type has been carried out by Sunderland et al. (1956) on the shoot apex of *Lupinus*. Protein content and rates of respiration in the apical meristem and subjacent leaf and stem tissues of the first seven internodes were examined in this study. It was found that, in general, protein concentration is higher in leaf primordia than in the associated stem segment but the respiration rate is lower. The apical dome gave values for both of these, which were intermediate between those of the first primordium and the first stem segment. Because cell volume is always lower in a leaf primordium than in the associated stem segment, it was concluded that the intermediate values obtained for the apical dome resulted from the presence of both kinds of cells. On this basis it was postulated that the shoot meristem is differentiated into two regions: a small-celled tunica with a high protein concentration and low metabolic activity, and a larger-celled corpus with low protein concentration and high metabolic activity. Metabolites produced in the corpus are presumed to be transferred to the tunica where they are essential to the maintenance of high cell division activity. It was suggested that the continuing differences between leaf primordium and stem segment had their origin in this primary differentiation of the meristem.

The suggestion of metabolic differentiation within the apical meristem is a valuable one, but it does not seem that this particular pattern based upon extrapolation from more mature regions of the shoot can be accepted uncritically in the absence of data for the meristem itself. It is particularly important to remember that this pattern does not correspond to any known cytological pattern in the meristem of *Lupinus*, and there is no evidence to suggest the kind of tissue specificity of meristem regions that is implied by the suggested pattern.

General comment

After reviewing the body of analytical work that has been carried out on shoot apices in the vascular plants, it no doubt would be desirable to conclude with a brief summary setting forth the main conclusions. Perhaps the best commentary upon this work is that it is not yet possible to do this in a meaningful way. The problems raised by the structural diversity that was considered in the previous chapter certainly have not been resolved by these analyses. Nonetheless it is a hopeful sign that increasing numbers of investigators are being motivated to supplement traditional histological methods by more precise, often quantitative, analyses in their search for a functional interpretation of the shoot apex.

Without doubt, the long-standing concept of an apical organization based upon the existence of permanent apical initial cells has been challenged from two different directions. On the one hand it is suggested that the true initiating region of the shoot is peripheral to the zone in which such cells would be found and that the central cells are essentially passive until the onset of reproductive development. On the other hand evidence is presented for a high degree of cell division activity in this region of living meristems, which then appears as a shifting population of cells without permanent entities that could be considered as apical initials in the classical sense. Between these extreme views is a third, according to which the summit of the apex is occupied by a group of distinctive cells that divide infrequently during vegetative development but frequently enough to function as the initiating center of the shoot, perhaps functioning in this way over extended periods of time because they do not divide frequently. This group of cells would seem to correspond to the traditional promeristem with its centrally placed apical initials. Over-all distinctive cytochemical patterns have been demonstrated in the meristem but their immediate bearing upon these alternative interpretations is far from clear. In short, this is a period of exciting activity in the study of shoot apical organization and function; and the conflicting evidence and differences of opinion that exist can be regarded only as manifestations of intense interest in this phase of plant development.

There is also another method of approaching the question of the functioning of the shoot apex, and this is by the medium of experimental manipulation of the apical region. It is appropriate to proceed in the next chapter to a consideration of some of this experimental work and its contribution to our understanding of the functional organization of the shoot apex.

REFERENCES

Ball, E. 1960. Cell divisions in living shoot apices. Phytomorphology **10**:377–396.

Brown, J. A. M., J. P. Miksche, and H. H. Smith. 1964. An analysis of H^3-thymidine distribution throughout the vegetative meristem of *Arabidopsis thaliana* (L.) Heynh. Radiation Botany **4**:107–114.

Buvat, R. 1952. Structure, évolution et functionnement du méristème apical de quelques dicotylédones. Ann. Sci. Nat. Bot. Ser. 11. **13**:199–300.

———. 1955. Le méristème apical de la tige. Ann. Biol. **31**:595–656.

Camefort, H. 1956. Étude de la structure du point végétatif et des variations phyllotaxiques chez quelques gymnospermes. Ann. Sci. Nat. Bot. Ser. 11. **17**:1–185.

Catesson, A. M. 1953. Structure, évolution et functionnement du point végétatif d'une monocotylédone: *Luzula pedmontana* Boiss. et Reut. (Joncacées) Ann. Sci. Nat. Bot. Ser. 11. **14**:253–291.

Clowes, F. A. L. 1961. Apical meristems. Blackwell, Oxford.

Dermen, H. 1945. The mechanism of colchicine-induced cytohistological changes in cranberry. Am. J. Botany **32**:387–394.

Esau, K., A. S. Foster, and E. M. Gifford, Jr. 1954. Mitosis in the initiation zone of the shoot apex. Congr. Intern. Botan., 8e, Rappt. Commun.: 262–263. Paris.

Gifford, E. M., Jr., S. Kupila, and S. Yamaguchi. 1963. Experiments in the application of H^3-thymidine and adenine-8-C^{14} to shoot tips. Phytomorphology **13**:14–22.

Gifford, E. M., Jr., and H. B. Tepper. 1962a. Histochemical and autoradiographic studies of floral induction in *Chenopodium album*. Am. J. Botany **49**:706–714.

———. 1962b. Ontogenetic and histochemical changes in the vegetative shoot tip of *Chenopdium album*. Am. J. Botany **49**:902–911.

Jacobs, W. P., and I. B. Morrow. 1961. A quantitative study of mitotic figures in relation to development in the apical meristem of vegetative shoots of *Coleus*. Develop. Biol. **3**:569–587.

Lance, A. 1952. Sur la structure et le functionnement du point végétatif de *Vicia faba* L. Ann. Sci. Nat. Bot. Ser. 11. **13**:301–339.

Newman, I. V. 1956. Pattern in meristems of vascular plants. I. Cell partition in living apices and in the cambial zone in relation to the concepts of initial cells and apical cells. Phytomorphology **6**:1–19.

Nougarède, A. 1967. Experimental cytology of the shoot apical cells during vegetative growth and flowering. Internat. Rev. Cytol. **21**:203–351.

Nougarède, A., R. Bronchart, G. Bernier, and P. Rondet. 1964. Comportement du méristème apical du *Perilla nankinensis* (Lour.) Decne. en relation avec les conditions photopériodiques. Rev. Gén. Botan. **71**:205–238.

Nougarède, A., E. M. Gifford, Jr., and P. Rondet. 1965. Cytological studies of the apical meristem of *Amaranthus retroflexus* under various photoperiodic regimes. Botan. Gaz. **126**:281–298.

Paolillo, D. J., and E. M. Gifford, Jr. 1961. Plastochronic changes and the concept of apical initials in *Ephedra altissima*. Am. J. Botany **48**:8–16.

Philipson, W. R. 1954. Organization of the shoot apex in dicotyledons. Phytomorphology **4**:70–75.

Popham, R. A. 1958. Cytogenesis and zonation in the shoot apex of *Chrysanthemum morifolium*. Am. J. Botany **45**:198–206.

Saint-Côme, R. 1966. Application des techniques histoautoradiographiques et des méthodes statistiques à l'étude du functionnement apical chez le *Coleus blumei* Benth. Rev. Gén. Botan. **73**:241–323.

Satina, S., A. F. Blakeslee, and A. G. Avery. 1940. Demonstration of the three germ layers in the shoot apex of *Datura* by means of induced polyploidy in periclinal chimeras. Am. J. Botany **27**:895–905.

Savelkoul, R. M. H. 1957. Distribution of mitotic activity within the shoot apex of *Elodea densa*. Am. J. Botany **44**:311–317.

Soma, K., and E. Ball. 1964. Studies of the surface growth of the shoot apex of *Lupinus albus*. Brookhaven Symp. Biol. **16**:13–45.

Steeves, T. A., M. A. Hicks, J. M. Naylor, and P. Rennie. 1969. Analytical studies on the shoot apex of *Helianthus annuus*. Can. J. Botany **47**:1367–1375.

Steward, F. C., R. H. Wetmore, and J. K. Pollard. 1955. The nitrogeneous components of the shoot apex of *Adiantum pedatum*. Am. J. Botany **42**:946–948.

Steward, F. C., R. H. Wetmore, J. F. Thompson, and J. P. Nitsch. 1954. A quantitative chromatographic study of nitrogeneous components of shoot apices. Am. J. Botany **41**:123–134.

Sunderland, N., J. K. Heyes, and R. Brown. 1956. Growth and metabolism in the shoot apex of *Lupinus albus*. *In* F. L. Milthorpe, [ed.] The Growth of Leaves p. 77–90. Butterworths, London.

Vanden Born, W. H. 1963. Histochemical studies of enzyme distribution in shoot tips of white spruce (*Picea glauca* [Moench] Voss.). Can. J. Botany **41**: 1509–1527.

Wardlaw, C. W. 1957. The reactivity of the apical meristem as ascertained by cytological and other techniques. New Phytol. **56**:221–229.

Wetmore, R. H., E. M. Gifford, Jr. and M. C. Green. 1959. Development of vegetative and floral buds. *In* R. B. Withrow, [ed.] Photoperiodism and related phenomena in plants and animals. Amer. Assn. Adv. Sci. Publ. **55**:255–273.

SIX

Experimental

Investigations

on the Shoot Apex

Most of the work reviewed thus far has consisted of observations of various types of shoot apices that are for the most part normal, or at least intact. It is also possible to investigate the organization of the meristem by subjecting it to a variety of experimental treatments and analysing its reaction to these manipulations. The most widely used and by and large the most successful of the experimental procedures applied to shoot apices has been microsurgery in which delicate instruments are used to make punctures, incisions, and excisions in various regions of the meristem. By this method it is possible to obtain information about the roles played by various portions of the meristem, and about the interactions between parts. One limitation of this method is that it cannot be assumed that the effects of operations are just the isolation or removal of particular regions of meristematic tissue. The unknown, and to a certain extent known, consequences of wounding may produce a variety of responses, such as cell proliferation, which are difficult to evaluate in terms of the normal apex. Thus surgical methods tell a great deal about the potentialities of portions of a meristem, information that is extremely valuable; but experiments must be controlled carefully and the results screened rigorously if the information is to be applied validly to the interpretation of the organization of the intact meristem.

Organization and integration
in the shoot apex

The structural and analytical evidence presented in earlier chapters suggests that there may be localizations of function within the shoot meristem. A variety of surgical methods may be used to investigate this possibility, the simplest being the separation of the meristem into two portions by a vertical incision. In a series of experiments that marked the beginning of the modern era of experimental investigation of shoot apices, Pilkington (1929) carried out operations of this type on young plants of *Vicia faba* and *Lupinus albus* (Fig. 6.1). One might anticipate several different results of such an operation, each of which would lead to a different

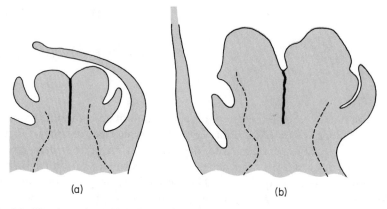

(a)   (b)

Fig. 6.1   The shoot tip of *Vicia faba* sectioned several days after the meristem had been bisected by a vertical incision. (a) Seven days after the operation the two halves of the meristem are enlarged. (b) Thirteen days after the operation new leaf primordia have emerged on the two regenerating apices. ×50. (M. Pilkington. 1929. New Phytol. **28**: 37.)

interpretation of the organization of the meristem. The apex might simply cease its growth or the two halves might graft together, reestablishing a single apex, suggesting that the apical meristem must be intact if it is to function at all. Alternatively the half apices might continue their development separately. In this event, each half meristem might give rise to what would correspond to one half of a normal shoot in its symmetry and in the tissues and organs that it contains. On the other hand each half meristem might become reorganized into a complete meristem capable of giving rise to a whole shoot with normal symmetry and placement of organs and tissues. Pilkington's experiments gave clear-cut evidence that this last alternative—the development of two complete shoots—is the correct one, and this finding has been confirmed by a number of subsequent workers. In a reexamination of this phenomenon in *Lupinus*, Ball (1955) has given considerable detail on the reestablishment of apical organization in the half meristems. In each half a new center was established on the original apical flank at some distance from the incision and not involving the original center. Thus it appears that a portion of the meristem, if separated, is capable of reestablishing a new meristem comparable in organization and function to the one from which it was separated.

The results of apical bisection raise the further question of whether smaller subdivisions of the meristem might be able to reestablish whole meristems. When the apex of *Lupinus* was separated into four segments (Ball, 1948) each of the quadrants regenerated a complete meristem and produced an entire shoot. When the apex was divided into six segments (Ball, 1952a) some of these regularly regenerated, but always fewer than the total number possible. The fact remains, however, that segments as small as one sixth of the original meristem were capable of reestablishing a complete meristem. When the meristem was divided into eight portions, none showed regenerative capacity. Subsequently Sussex (1952) was able to obtain complete regeneration from a panel on the flank of the meristem of *Solanum tuberosum* (potato), which represented only about one twentieth of the original meristem (Fig. 6.2). However, this regeneration occurred only if the rest of the meristem was excised from the plant, leaving the panel, with about 12 cells in its surface layer, alone at the summit of the shoot. Such small pieces of meristem pass through an initial period during which they enlarge considerably before initiating any leaf primordia. It must be concluded that the minimal size of a piece of meristem that can regenerate completely is very small and that minimal size has not yet been defined. In fact one might ask whether, if the individual cells of the meristem could be isolated, they would reestablish whole meristems.

These experiments show that small portions of the terminal meristem have the capacity to regenerate entire apices; but only Loiseau (1959) seems to have tested the possibility that there may be localization of function within the meristem by the isolation of specific regions. This worker destroyed large areas of the apical surface, leaving relatively small groups of cells either in the central (méristème d'attente) or in the peripheral (anneau initial) regions. The resulting development indicated an equivalent ability to regenerate in the two regions.

In the study of developmental potentialities of precise regions in the animal embryo, it has been profitable to use the technique of grafting in which a particular organ rudiment or tissue mass is excised and is implanted in another location. The possibility that comparable experiments

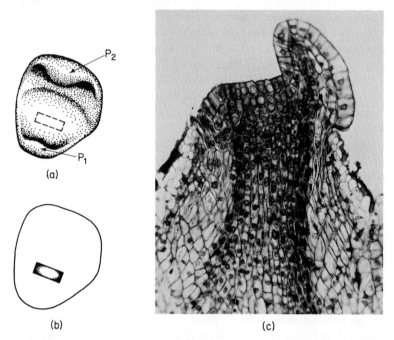

Fig. 6.2   Regeneration of the terminal meristem of the shoot of *Solanum tuberosum.* (a) A diagram of the shoot apex showing the two youngest leaf primordia, P_1 and P_2, and the position of the apical panel that will be tested in the experiment. (b) The shoot apex after excision of all meristematic tissue except that in the apical panel. (c) A regenerating apical panel that has formed a leaf primordium and contains procambium. (c) ×94. ((b, c) I. M. Sussex. 1964. Brookhaven Symp. Biol. **16**: 1.)

might be carried out in the analysis of the shoot meristem of plants was explored by Ball (1950) in *Lupinus*. He cut wedge-shaped pieces from the summit of the apex and attempted to reimplant them in the original position as a preliminary test, and to exchange such pieces between comparable apices. Unfortunately this promising approach had to be abondoned because even in the few instances in which the grafted piece remained alive for some time, no growth was detected.

Although it appears that any portion of the shoot meristem, even a relatively small portion, has the capacity to regenerate an entire apex, the fact remains that in the intact apex this potentiality is not expressed. Clearly there is a mechanism that serves to maintain the functional integrity of the shoot meristem. Two of the possible mechanisms that might be suggested are hormonal integration and competition for nutrients among various parts of the apex. It has been shown, particularly in the case of the isolation of small portions of meristem (Sussex, 1953), that shallow incisions not penetrating beyond the upper four or five layers of the apex are considerably less effective in permitting the regeneration of isolated panels than are deep incisions extending into the maturing subapical region. Such results are suggestive of a competetive nutritional basis of the integration mechanism. On the other hand, many plant physiologists would find it impossible to consider distribution of nutrients without invoking a hormonal mediation of this process.

Integration of the meristem has also been explored by surgical experiments of another type in which the distal cells of the apex are destroyed by puncturing with a needle. In *Lupinus albus* and *Vicia faba* Pilkington (1929) reported that puncturing the summit of the apex was invariably followed by regeneration of only one new apex from the flank region of the original meristem. Sussex (1964) observed a similar response to needle punctures of varying diameters in the shoot apex of *Solanum* (Fig. 6.3),

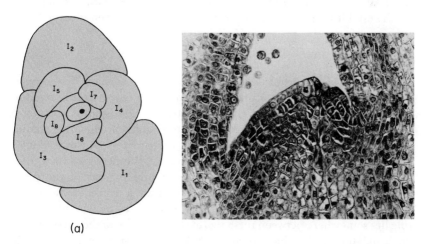

(a)

Fig. 6.3   Apical puncture in *Solanum tuberosum*. (a) Diagram of a shoot apex in which eight leaves (I_1–I_8) had been initiated while the puncture remained in a central position in the meristem. (b) A longitudinal section of an apex showing the terminal puncture. (b) ×87. (I. M. Sussex. 1964. Brookhaven Symp. Biol. **16**: 1.)

but a different result was apparently obtained in *Impatiens* by Loiseau (1959). After punctures had been made in the summit there was a general expansion of the apex followed by reorganization into two, three, or four new shoot apices. When the same experiment was performed on the fern *Osmunda cinnamomea*, two to six new apices regenerated from the uninjured portions of the meristem. In this case the original puncture was small, including only the apical cell and a few adjacent derivatives; but subsequent necrosis produced an effect corresponding to an extensive punc-ture. In the fern *Dryopteris* Wardlaw (1949) observed that similar punc-tures led to the regeneration of one or more buds from the peripheral parts of the original meristem.

Though experiments such as these do not indicate the mechanism by which integration in the meristem is achieved, they do suggest that the central group of cells may be especially important in this phenomenon. In all of these cases, the destruction of these cells has been followed by a reorganization of the meristem such that one or more new growth centers have been established from uninjured portions of the meristem. Equivalent injuries in peripheral regions (Loiseau, 1959) have no such effect and, in fact, are soon displaced by further growth of the apex along its original axis. In plants that branch terminally by separation of the meristem into two equal or unequal parts, something akin to inactivation of the summit cells seems to occur. In both *Selaginella willdenovii* and *Osmunda cinnamomea* the loss of apical meristem characteristics by these cells, accompanied by the establishment of two new growth centers, has been observed.

Apical autonomy

One of the problems confronting students of plant development is the question of the relationship between the meristem and the mature regions of the plant it has produced. Clearly the meristem is a center of cell production and it provides the structural units of which the mature body is constructed. It is not, however, immediately obvious whether the meristem functions in a completely plastic manner, governed in all of its activities by the patterns of more mature regions of the plant, whether it is largely autonomous, perhaps to a considerable extent controlling the destinies of its own cellular products, or whether its true role lies somewhere between these two extremes. Some of the experimental work on shoot apices has provided evidence favoring the view that the meristem is highly autonomous in its activity.

In both ferns and angiosperms (Wardlaw, 1950; Ball, 1952b) the center of the apical meristem has been isolated inside the youngest leaf primordia by three or four deep, vertical incisions that leave it supported on a plug of maturing pith tissue and separated laterally from all leaf primordia (Fig. 6.4). Its normal vascular supply is, of course, severed by the incisions. Under these conditions the meristem continues its growth, although often at a reduced rate at the outset, initiates a sequence of new leaf primordia, and gives rise to normal tissues of the shoot. Vascular differentiation, furthermore, proceeds basipetally through the original immature pith tissue and often establishes a connection with the original shoot vascular system around the bases of the incisions. The seemingly

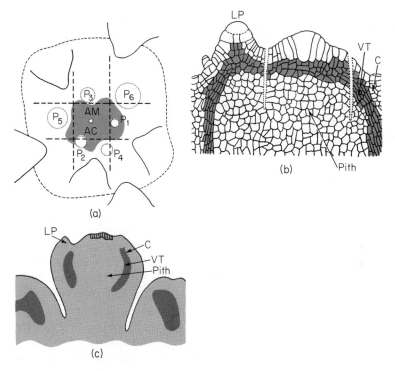

Fig. 6.4   Surgical isolation of the apical meristem of *Dryopteris dilatata*. (a) A view of the shoot tip showing the position of the four incisions that isolated the terminal meristem from the surrounding leaf primordia and differentiating tissues. (b) Longitudinal section of a surgically isolated meristem showing the tissues which have been severed by the incisions, and the intact pith connection below the meristem. (c) A longitudinal section of the shoot tip five weeks after an operation isolating the meristem. The meristem has continued to grow, producing new leaf primordia; and cortical, vascular, and pith tissues have been differentiated internally. Key: AC, apical cell; AM, apical meristem; C, cortex; LP, leaf primordium; P_1–P_6, the six youngest leaf primordia around the meristem; VT, vascular tissue. (a) ×30, (b) ×70, (c) ×15. (Adapted from C. W. Wardlaw. 1947. Phil. Trans. Roy. Soc., London, B **232**: 343.)

normal development of the meristem under these conditions makes it difficult to believe that it has any dependence upon mature regions of the plant for anything other than basic nutrients.

With the development of sterile culture techniques it was possible to effect an even more dramatic isolation of the shoot meristem. Among the early studies of this type, that of Ball (1946) is particularly relevant. In this investigation pieces from the shoot tip approximately 0.5 cubic millimeters in volume and including the shoot meristem and several of the youngest leaf primordia were explanted to a culture medium containing mineral salts, sugar, agar, and water. Such explants regularly gave rise to complete shoots and, through rooting, to whole plants that could subsequently be transplanted to a greenhouse. Results of this sort, though they are suggestive of apical autonomy, do not test the potentialities of the meristem alone because several leaf primordia and considerable subjacent tissue were included in the explant. Subsequently Ball (1960) was able to grow meristems of *Lupinus* isolated without any leaf primordia into shoots five to ten centimeters in length and bearing seven to nine leaves. This result was achieved only by the incorporation of coconut milk and gibberellic acid into the medium; and shoot growth was ultimately terminated by the loss of meristematic characteristics by the meristem.

The failure to obtain unlimited growth from the isolated meristem of an angiosperm even on a complex medium was perplexing since similar explants from several species of vascular cryptogams had been shown to be able to continue their growth without limitation on a medium of mineral salts and sugar (Wetmore, 1954). They ultimately gave rise to whole plants (Figs. 6.5, 6.6). More recently this difficulty has been resolved by Smith and Murashige (1970) who have been able to obtain whole plants from explants consisting only of the terminal meristem without leaf primordia in five species of flowering plants. The medium required for this development was remarkably simple, containing in addition to mineral salts and sugar only myo-inositol, thiamin hydrochloride, and indoleacetic acid. It was established that indoleacetic acid was essential for the complete development of the meristem. The pattern of development of the isolated meristems was followed in detail in only one species, tobacco (*Nicotiana tabacum*), and it differed strikingly from that of larger explants cultured earlier in which a shoot formed first and subsequently initiated roots.

Fig. 6.5   Longitudinal section of the shoot tip of *Adiantum* showing the size of the meristem piece that was excised and grown in sterile culture.

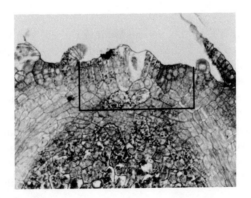

Fig. 6.6   A plant of *Adiantum* grown in culture from an explant like that shown in Fig. 6.5. (R. H. Wetmore. 1954. Brookhaven Symp. Biol. **6** : 22.)

During the first six days in culture the meristem underwent considerable cell division and cell enlargement, particularly near the cut surface, but it retained its essential organization. In six to nine days a root apex was initiated within the tissue opposite the continuing shoot apex so that the resulting axis was bipolar and showed a remarkable resemblance to an embryo except for the absence of cotyledons. By the twelfth day a leaf primordium was initiated and the shoot-root axis contained a well developed provascular system. Subsequent growth led to the development of a plantlet which ultimately could be transplanted to soil.

Thus the culture experiments, like the surgical studies, indicate a high degree of independence or autonomy in the meristem. The long debated difficulty in culturing angiosperm apices seems to have resulted from a failure to find the precise nutrients and growth factors required for normal development, and it is probable that these differ in different species. However, one may reasonably ask whether an organ which is dependent upon exogenous sources of such specific substances as indoleacetic acid can be regarded as truly autonomous in any ordinary sense of the word. Moreover, there are well-known phenomena in plants in which the shoot meristem responds dramatically to a precise exogenous stimulus. Processes such as the initiation of flowering, shoot tip abortion, and other similar developmental responses clearly indicate that the meristem may be extremely sensitive to stimuli proceeding from more mature regions of the plant. One aspect that must be kept in mind when dealing with apical autonomy is the distinction between those exogenous factors that simply may permit the meristem to regulate its own development and those that specifically induce new patterns of development.

General comment

The experiments that have been described in this chapter provide useful information about the functional organization of the shoot apex, but they do not offer final answers to the questions raised by the structural and analytical studies previously considered. The apex has been shown to be complex in its organization and not to be a homogeneous group of dividing cells. Yet subdivision of the apex into small units does not reveal localized differences in the ability to regenerate an entire shoot apex. On the other hand, in the normal, unoperated apex all of these cells function together as an integrated system, indicating that there must be a regulating mechanism that suppresses the tendency for regeneration of the various parts and maintains organization of the apex as a whole. Experimental evidence showing that surgical disruption of this organization depends upon deep cuts has suggested that competition for nutrients between different parts of the apex may be a factor in maintaining the organization. The terminal meristem, however, possesses a high degree of autonomy in its ability to develop normally after isolation or excision; and this favors the conclusion that the basis of organization, whether nutritional, or hormonal, or a combination of the two, is located within the apex itself.

It would be most rewarding to be able to conclude this discussion of the shoot apex by offering an interpretation of the organization of this region, and particularly by relating this to the structural patterns that have been described. Unfortunately it would be premature to attempt any extensive interpretation along these lines at this time. Some surgical evidence points to the central cells at the summit of the apex as being particularly significant in maintaining the integrity of the terminal meristem, even though these cells do not show any more regenerative potency than do those more peripherally placed. The significance of the summit cells is most obvious in those lower vascular plants where a distinct apical cell is found. Such evidence might be taken as giving at least some support to the concept of apical initials, or to a special regulatory function of the central zone. Similarly, the equal regenerative ability of all parts of the terminal meristem might lend support to the concept of an undifferentiated promeristem. Until a clearer understanding of the mechanism of integration in the meristem is formulated, however, it would be unwise to attempt such generalizations.

If the question of apical organization is considered from another point of view, it is difficult to avoid the conclusion that the structural complexity of the meristem—within which can be discerned zonation patterns, localized differences in cell division frequency, and different orientations of cell divisions in surface and interior layers—reflects a delicate balance maintained by refined controls. There are experiments, as yet rather difficult to interpret consistently, which show that the direct application of growth-regulating substances to the terminal meristem can, without actual surgery, upset this balance and thus change both the patterns and the functional organization of the meristem. Similarly, nonlethal doses of ionizing radiations can cause changes in planes of cell division, and have also revealed differences in sensitivity of different regions of the apex. As more is learned about the biochemistry of the shoot apex, such methods may be expected to become increasingly useful in exploring the nature of the mechanisms that maintain organization in this formative region. One of the most interesting, and potentially most useful, interpretations of the organization of the shoot apex views the apex as dominated by a system of fields of physiological activity associated with growth centers that consist of the apex itself and its recently produced leaf primordia. Because a discussion of the field theory of necessity must involve questions of organ initiation and development, examination of this approach to apical organization will be deferred to a later chapter dealing with leaf and bud initiation.

REFERENCES

Ball, E. 1946. Development in sterile culture of stem tips and subjacent regions of *Tropaeolum majus* L. and of *Lupinus albus*. L. Am. J. Botany **33**:301–318.

———. 1948. Differentiation in the primary shoots of *Lupinus albus* L. and of *Tropaeolum majus* L. Symp. Soc. Exp. Biol. **2**:246–262.

———. 1950. Isolation, removal and attempted transplants of the central portion of the shoot apex of *Lupinus albus* L. Am. J. Botany **37**:117–136.

———. 1952a. Experimental division of the shoot apex of *Lupinus albus* L. Growth **16**:151–174.

———. 1952b. Morphogenesis of shoots after isolation of the shoot apex of *Lupinus albus*. Am. J. Botany **39**:167–191.

———. 1955. On certain gradients in the shoot tip of *Lupinus albus*. Am. J. Botany **42**:509–521.

———. 1960. Sterile culture of the shoot apex of *Lupinus albus*. Growth **24**:91–110.

Loiseau, J. E. 1959. Observation et expérimentation sur la phyllotaxie et le fonctionnement du sommet végétatif chez quelques Balsaminacées. Ann. Sci. Nat. Bot. Ser. 11 **20**:1–214.

Pilkington, M. 1929. The regeneration of the stem apex. New Phytol. **28**:37–53.

Smith, R. H., and T. Murashige. 1970. *In vitro* development of the isolated shoot apical meristem of angiosperms. Am. J. Botany **57**:562–568.

Sussex, I. M. 1952. Regeneration of the potato shoot apex. Nature **170**:755–757.

———. 1953. Regeneration of the potato shoot apex. Nature **171**:224–225.

———. 1964. The permanence of meristems: Developmental organizers or reactors to exogenous stimuli? Brookhaven Symp. Biol. **16**:1–12.

Wardlaw, C. W. 1949. Further experimental observations on the shoot apex of *Dryopteris aristata Druce*. Phil. Trans. Roy. Soc. London Ser. **B 233**:415–451.

———. 1950. The comparative investigation of apices of vascular plants by experimental methods. Phil. Trans. Roy. Soc. London Ser. **B 234**:583–604.

Wetmore, R. H. 1954. The use of *in vitro* cultures in the investigation of growth and differentiation in vascular plants. Brookhaven Symp. Biol. **6**:22–40.

SEVEN

Organogenesis in the Shoot—Leaf Origin and Position

Any attempt to study the apical meristem of the shoot, even a simple dissection to expose it for observation, leads immediately to a consideration of the initiation and development of lateral appendages, particularly leaf primordia. It is evident that the formation of leaf primordia is a major activity of the shoot meristem and that the early development of these primordia in such close proximity to the meristem must result in important developmental interactions between the two. This intimate relationship is reflected in the unity of the mature shoot system in which any attempt to isolate stem and leaf either structurally or functionally is artificial. The student of phylogeny finds an interpretation of this relationship, at least in the ferns and seed plants, in the evolution of stem and leaf from a primitively undifferentiated branching system. The object of developmental analysis, however, must be to understand how, in the individual living plant, structures that originate as outgrowths of the meristem acquire distinctive characteristics and interact with one another and with the meristem that produced them. This analysis is further complicated by the fact that the same meristem frequently gives rise to other appendages that develop as replicas of the original shoot.

Whereas the shoot is characterized by potentially unlimited or indeterminate growth and this feature is retained by lateral branches, the leaf is an organ of transient, although in some cases extensive, growth. Thus the leaf at its inception has a definable developmental destiny, which is to produce a lateral appendage mature in all its parts and devoid of further growth potentialities. This is equally true for the minute scales of a juniper or for the massive fronds of a tree fern and it may be correlated with the fundamental role of the leaf as a photosynthetic organ with a limited functional life. Often the leaf is further distinguished by its dorsiventrality and elaborate structural specialization. In developmental terms, the problem is to account for the change in growth pattern of a particular group of cells of the apical meristem that results in the formation of an organ as distinctive as a leaf. Moreover, because this change is localized and occurs only in precise positions in the shoot meristem, it is necessary to seek the mechanisms that regulate this activity of the meristem. In many ways the origin and development of a determinate organ such as the leaf resemble the formation of appendages in some animals and may provide a better opportunity to discover similarities in plant and animal development than that provided by the indeterminate activity of the apical meristem, which is without an obvious equivalent in animals.

Leaf initiation

Perhaps the best way to approach the problem of leaf initiation and other aspects of early leaf development is to consider a specific instance in which a detailed analysis has been carried out. The potato (*Solanum tuberosum*) will serve as an example (Sussex, 1955). The first suggestion of leaf initiation is the swelling of the apical flank in the position that may be predicted to be the site of the next leaf primordium (Fig. 7.1a). This protrusion is the result of accelerated cell division and growth in at least three layers of the meristem and probably deeper tissues as well. Because potato has a two-layered tunica, both tunica and corpus are involved in the initiation process. In the surface layer only anticlinal divisions occur, but in the underlying layers both anticlinal and periclinal divisions may be observed. The protruding flank of the apex which precedes the emergence of a distinct primordium has been designated the *foliar buttress* in many species of dicotyledons.

The emergence of the leaf primordium from the foliar buttress is accomplished by the localization of growth activity in a central group of cells that, in effect, becomes the apical meristem of the leaf (Fig. 7.1b). In many dicotyledons it is possible to recognize a conspicuous cell beneath the surface protoderm that has been designated the *subapical initial* and assigned a special initiating role in leaf apical growth. In potato such a cell can be recognized. This cell divides both periclinally and anticlinally and thus initiates much of the internal tissue of the leaf. In other dicotyledons there is no single distinctive cell in this position; and it is ques-

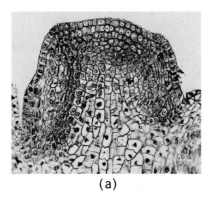
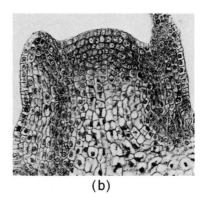

(a)   (b)

Fig 7.1   Stages in the emergence of a leaf primordium of *Solanum tuberosum*. (a) A well-developed leaf buttress is formed on the left flank of the meristem. Below it procambium can be seen in the stem. On the right flank of the meristem can be seen the edge of an older leaf. (b) A slightly older stage of leaf emergence. The new leaf appears on the left side of the meristem as a protrusion at the summit of the buttress. ×69. (I. M. Sussex. 1955. Phytomorphology **5** : 253.)

tionable, therefore, whether particular significance should be attached to any one cell at the leaf apex. In potato, growth at the leaf apex continues until four younger primordia have been initiated and a length of approximately 200 microns has been achieved. At this point apical growth ceases and the terminal leaf cells become vacuolated. There is great variation among dicotyledons in the extent of apical growth, but in the great majority thus far investigated this process ceases before the primordium has attained a length of one millimeter. There are, however, notable exceptions such as *Nicotiana tabacum* (tobacco) where leaf apical growth continues until the primordium reaches a length of three millimeters and *Archangelica officinalis* with a length of 15 millimeters. The important point is that after a definite period growth at the apex ceases so that, in contrast to the shoot as a whole, the leaf is determinate.

In potato the newly emergent primordium is essentially circular in section. As growth continues a progressive change in shape occurs in the primordium, and it becomes somewhat flattened on the side facing the shoot meristem (Fig. 7.2). This change in shape results from more extensive growth throughout the primordium in the tangential dimension than in the radial. Thus by the end of the apical growth phase the primordium consists of a bilaterally symmetrical, peg-like outgrowth.

Whereas the initial manifestation of bilaterality in the primordium results from the growth pattern of the whole organ, the extension of this in later development is brought about by localized growth along the

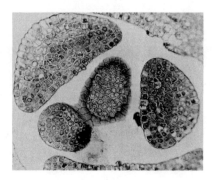

Fig. 7.2   A cross section of the shoot tip of *Solanum tuberosum* showing the shape changes of developing leaf primordia. The youngest primordium, shown attached to the shoot apex at the lower left, is approximately circular in section, the two older leaves are increasingly bilateral as a result of growth of the marginal meristems. ×60. (I. M. Sussex. 1955. Phytomorphology **5**: 253.)

leaf margins. Marginal growth is ordinarily treated as a distinct phase of leaf development; but because it is usually described as beginning before the end of apical growth, it is clear that apical and marginal phases overlap. Moreover, the pattern of meristematic activity in the two regions is sufficiently similar to suggest that marginal activity may in fact represent a displacement to a lateral position of apical development. The particulars of marginal meristem activity will be described in a later chapter in connection with histogenesis in the leaf.

Descriptions of leaf initiation and early development in other species indicate that there is considerable variation in the size relationship between a leaf primordium and the meristem that produces it. If the primordium is sufficiently large that its initiation involves a sizable portion of the meristem, its initial stage is evident as a lateral bulge of the meristem, that is, a leaf buttress. If, however, the primordium is small in relation to the meristem as in the high-domed apices of grasses and some other monocotyledons or the broad conical apices of many ferns, the first emergence is ordinarily recognizable as a distinct leaf primordium and no buttress stage is designated. Published accounts of leaf initiation reveal the variations in size relationships between apical meristem and leaf primordium; but the accounts themselves vary in the recognition of a buttress stage. The foliar buttress, where it occurs, is the leaf in its earliest stage of emergence. On the other hand many workers have considered buttress formation as a phase of apical expansion preceding leaf initiation and have described a sequence of maximal and minimal sizes of the meristem preceding and following successive leaf initiations. Indeed, the buttress is both of these things, a fact that emphasizes the difficulty of separating stem and leaf in the unified shoot system. This question is more than a matter of terminology because the very early stages of development must be identified precisely and described clearly as a basis for experimental work on leaf initiation. The importance of this point will be evident in the interpretation of experimental studies that will follow.

In the study of leaf initiation, a great deal of attention has been devoted to establishing the precise depth of tissues in the apical meristem from which the primordium arises, particularly in the stratified apices of flowering plants. Although it is relatively easy to locate the site of early cell divisions associated with leaf initiation, it is much more difficult to determine which apical layers contribute tissues to the total development of the leaf. The only completely effective means of tracing meristem contributions to the leaf has been the study of periclinal chimeras in which meristem layers are marked by their cellular characteristics and may thus be traced into the developing leaf. The most useful of these have been the polyploid chimeras discussed in Chapter Five. In both *Datura* and *Oxycoccus* (cranberry) the leaf appears to be composed of tissues derived from the three outermost layers of the meristem, thus including both tunica and corpus. Thus in the two cases in which reliable evidence is available it appears that leaf formation is not relegated simply to superficial regions of the meristem but involves its full depth. As further evidence is obtained, it will be interesting to learn whether this is a general principle throughout the angiosperms and gymnosperms.

On the other hand, in the lower vascular plants where the apical meristem consists of a single layer of cells, it might be expected that the relationship between leaf primordium and meristem would be different, at least in details if not in principle. Because much of the experimental work on leaf initiation has been carried out on several species of ferns, it will be advisable to consider early leaf development in such a plant. In *Dryopteris* (Wardlaw, 1949b) and in *Osmunda* (Steeves and Briggs, 1958) a group of cells at the periphery of the surface apical layer initiates the leaf by an acceleration of cell division activity (Fig. 7.3a). Wardlaw has described an initial stage in which the cells of this group enlarge; but no such enlargement has been noted in *Osmunda*. Accelerated growth soon leads to the protrusion of a primordium that is roughly circular in outline and, being small in relation to the total volume of the apex, is immediately distinct as a leaf primordium. At an early stage a centrally placed superficial cell of the primordium enlarges and becomes the distinctive apical initial cell of the leaf (Fig. 7.3b). In contrast to the situation in seed plants, apical growth is of long duration in fern leaves; and the apical cell remains visible and active for nearly three years in *Osmunda*. Undoubtedly the circinate or coiled pattern of the unexpanded

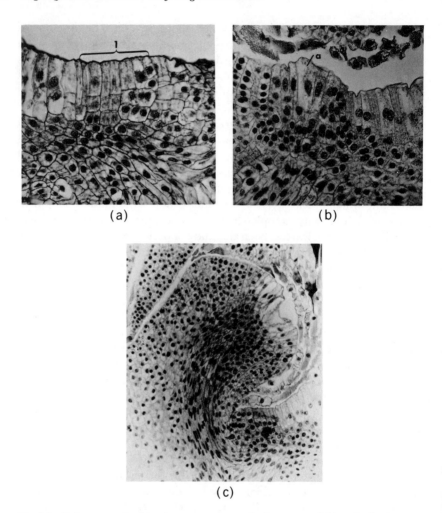

Fig. 7.3   Early stages of leaf development of *Osmunda cinnamomea.* (a) Longitudinal section of a very young leaf. Although it has not yet emerged as a mound on the surface of the shoot, active development in the form of cell division has taken place in the subsurface layers. (b) A slightly older stage in which the primordium is beginning to form a mound, and its tip now contains an enlarged apical cell. (c) A later stage of leaf growth. The leaf curves toward the center of the shoot apex, and the larger size of cells on the outer side of the primordium is evident.

frond axis is correlated with this long continuation of true apical growth. Ultimately, however, with one or two possible exceptions, the determinate nature of the organ is manifested in a cessation of apical growth and the disappearance of the apical cell. As in the seed plants, bilaterality is established at an early stage and is accentuated by the tendency of the primordium to curve towards the center of the meristem as a result of greater longitudinal growth on the abaxial side than on the adaxial (Fig. 7.3c).

The initiation and early development of leaves as they have been reviewed here raise important, and probably complex, questions concerning the regulation of development in plants. The leaf appears to be an unusually favorable system in which to analyze developmental processes because it is an easily recognizable entity and one that is readily accessible to the experimenter. It is not surprising, therefore, that a great deal of experimental work has been done on this system. The body of experimental data resulting from these investigations presents a complex and sometimes confusing picture that must be examined closely if its meaning is to be grasped. In order to facilitate this examination it is advisable to state at the outset what seem to be the essential questions posed by the phenomenon of leaf development. These questions may be framed as follows:

1. Why does the peripheral region of the shoot meristem give rise to outgrowths of any sort?
2. By what mechanism is the placement of the outgrowths regulated in the meristem?
3. What is the nature of the influences that act upon the outgrowths in such a way as to cause them to be leaves?
4. What is the nature of the response to these influences that results in leaf development?

In many ways, the first of these questions is the most fundamental, yet it is the aspect of leaf development about which the least information exists. One might postulate an inherent tendency of the meristem to grow out, at least in its peripheral regions. In fact Schüepp (1917) formulated a theory of leaf initiation for flowering plants based upon the idea that the anticlinally dividing surface layers should increase in area disproportionately to the irregularly dividing inner regions and thus be thrown into folds. The surgical work reviewed in the previous chapter has shown that, whereas the intact meristem functions as an integrated unit, small portions isolated laterally will grow out independently. The fact that such outgrowths give rise to shoots has no bearing upon the present question. Thus it is possible to visualize the initial phase of leaf development as resulting from an isolation of particular regions of the meristem from the correlative influences that otherwise prevent outgrowth. Unfortunately there is no information whatsoever as to the nature of the correlative mechanism or the system by which it is selectively interrupted, or, for that matter, whether such a visualization has any validity.

Leaf position—phyllotaxy

Descriptive account

The importance of the second question is evident even from observations on mature shoots where, although patterns vary, leaves are always distributed along the stem in a definite and repetitive manner. The arrangement of leaves on an axis is called *phyllotaxy*. Leaves may occur singly at each node, in pairs, or in whorls of three or more. Where they are individually disposed along the stem, it has long been recognized that they form a helical pattern ascending the stem in order of decreasing age. In plants with helical phyllotaxy it is also possible to recognize approximately vertical ranks of leaves, each of which is called an *orthostichy*. Customarily the phyllotatic pattern is described in the form of a fraction, the denominator of which is the number of orthostichies in the system and the numerator of which is the number of gyres of the helix between successive leaves on the same orthostichy.

In shoots in which there is little or no internodal elongation during maturation, phyllotaxy cannot be determined in this way because vertical ranks of leaves cannot be found. This can be seen very easily in a pine cone or in the head of a sunflower. This situation is of special interest to the student of development because it prevails generally in the unexpanded apical bud even though the mature shoot derived from the bud may have evident orthostichies. There is, however, another method of establishing the phyllotactic pattern in such cases; and this method may be described best by using an example. In *Osmunda cinnamomea*, if a transverse section passing through the apical meristem is drawn or projected in such a way that the positions of the youngest primordia and the bases or traces of older primordia are accurately placed, a different kind of pattern may be detected (Fig. 7.4). Again there are no evident straight ranks, but it is possible to observe curved rows of leaves extending outward from the center. These curved rows, which have been shown to have the path of logarithmic spirals, are called *parastichies*.

The parastichies that have been of greatest interest in the study of development are those known as contact parastichies of which there are two sets, one in each direction. Each leaf is located at the intersection of two contact parastichies and the leaves most closely associated with it on the two parastichies are those closest to it in the bud and even in contact with it in cases in which close packing occurs. The leaf interval along parastichies is regular, but the two sets of parastichies in opposite directions have different characteristics. Those of one set are more numerous, have a larger leaf interval, and are shorter than those of the other set.

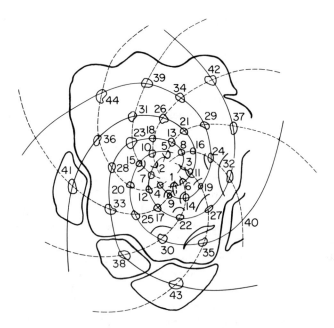

Fig. 7.4 A transverse section of the shoot tip of *Osmunda cinnamomea* with the contact parastichies drawn in. There are two sets of parastichies. One set, represented by continuous lines in the diagram, runs clockwise from older to younger leaves, and connects leaves differing in age by five plastochrons. There are five parastichy lines in this set. The other set, indicated by dashed lines, runs counterclockwise and connects leaves that differ in age by eight plastochrons. There are eight parastichy lines in this set. Thus the phyllotaxy is 5+8. (T. A. Steeves. 1963. J. Indian Bot. Soc. **42A** : 225.)

Phyllotaxy can be expressed in terms of the parastichies by stating the number in each set (Church, 1904). Thus in *Osmunda* it is possible to recognizes five long parastiches and eight short parastichies; and the phyllotaxy may be indicated as five plus eight contact parastichies. It is also possible to translate phyllotaxy expressed in this way to a fractional form, by taking the number of long parastichies as the numerator of the fraction and the total number as the denominator.

The apparent anomaly that an apical bud that contains no straight ranks of leaves should give rise to a shoot with orthostichies may be explained by a consideration of events during elongation. In the apical bud, in addition to the contact parastichies, there are other less sharply curved rows of leaves that, although they are not straight, correspond to orthostichies in the particular system in terms of the leaf interval along them. When elongation of the stem occurs and the leaves are widely separated, the lateral displacement of successive leaves in the row is minimized and the vertical alignment is accentuated so that the eye is caught by an apparent straight line.

Apart from plants having whorled or opposite phyllotaxy, the vast majority of vascular plants whose phyllotaxy has been described falls into a series that has interesting mathematical properties. In this series, the numbers of contact parastichies in the two sets are consecutive numbers in the Fibonacci series 1: 1: 2: 3: 5: 8 and so on, for example, 2 plus 3, 3 plus 5, 5 plus 8. The regularity of pattern revealed by the study of parastichies ultimately reflects the regularity of leaf initiation and particularly the fact that each new leaf primordium arises at a constant angular divergence from the next older one. If the divergence angles between successive leaves are measured in any plant, it will be found that, though fluctuations occur, the average is ordinarily an approximation of the so-called "ideal angle," 137°30′28″. Moreover it will be noted that if phyllotaxy is expressed in fractional form as just explained, successive fractions in the series, taken as fractions of 360°, converge on this angle. The important point is that, as far as can be determined by measurement, the angle of divergence remains constant whereas phyllotactic patterns expressed in terms of parastichy numbers vary widely.

It is thus a matter of some interest to consider how the phyllotactic patterns can vary while the divergence angle remains constant. Examination of apical sections of a plant having a low phyllotaxy (2 plus 3) and of a plant having a higher phyllotaxy (3 plus 5) indicate how this difference may be explained (Fig. 7.5). The phyllotactic pattern is determined by the number of contact parastichies in each direction and thus, ultimately, by the position in the total leaf sequence of the two leaves that are the contacts for any particular primordium. In the 2 plus 3 pattern the contact leaves for primordium 1 are leaves 3 and 4, whereas in the 3 plus 5 pattern the corresponding contacts are leaves 4 and 6. Variations in contact relationships among the leaves of an apical bud would seem to have their explanation in the growth relationships of apical meristem and leaf primordia. This is illustrated in Fig. 7.6 in which the branches of different size of *Araucaria excelsa* show different phyllotactic patterns resulting not from variations in the angle of divergence but from different relative growth rates of apex and leaf primordia. Richards (1951) has called attention to this fact in his use of the *plastochron ratio* in the interpretation of phyllotactic patterns. Briefly, this ratio is obtained by dividing the distance from the center of the apex to the center of one primordium by the distance from the apical center to the center of the next younger primordium; it thus gives a measure of the extent of radial growth by the apex during a plastochron, that is between the initiation of successive leaf primordia. The role of this variable in different phyllotactic patterns

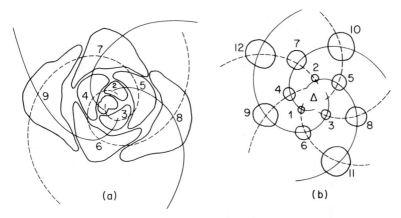

Fig. 7.5   Phyllotaxy of *Solanum tuberosum* (left) and of *Dryopteris dilatata* (right). In the potato shoot there are two parastichies in the clockwise direction (dashed lines) and three in the counterclockwise direction (continuous lines). The phyllotaxy is 2+3. Leaves are large in relation to the size of the shoot meristem at the time of their initiation, and they grow rapidly. In the *Dryopteris* shoot there are three counterclockwise parastichies and five clockwise parastichies. The phyllotaxy is 3+5. Leaves are small in relation to the size of the shoot meristem at the time of their initiation, and their rate of lateral expansion is slow.

has been demonstrated by Richards both in theoretical diagrams and in actual measurements in shoots with changing phyllotaxy. The primordial contacts in the apex, and consequently the pattern of contact parastichies, also are influenced by the shape of the developing primordia following their initiation. Because neither plastochron ratio nor primordial shape has any influence upon angular divergence, this factor remains constant in varying phyllotactic patterns.

Experimental Studies

The foregoing analysis of phyllotaxy is a purely descriptive account of the regular placement of leaves on the axis and provides little information about the mechanism that underlies the pattern. For this kind of understanding it has been necessary to turn to an experimental approach, the most successful technique having been a surgical one. Experiments of this type first were performed by M. and R. Snow (1931) on the shoot apex of *Lupinus* and led to the conclusion that the position at which a new leaf primordium is initiated is influenced by the preexisting leaf primordia adjacent to the site of initiation. If P_1, the youngest leaf primordium, was isolated from the apex by a tangential incision, the next leaf, I_1, arose in its expected position, but the following leaf, I_2, did not arise in its anticipated position (Fig. 7.7a, b). It was, in fact, abnormally

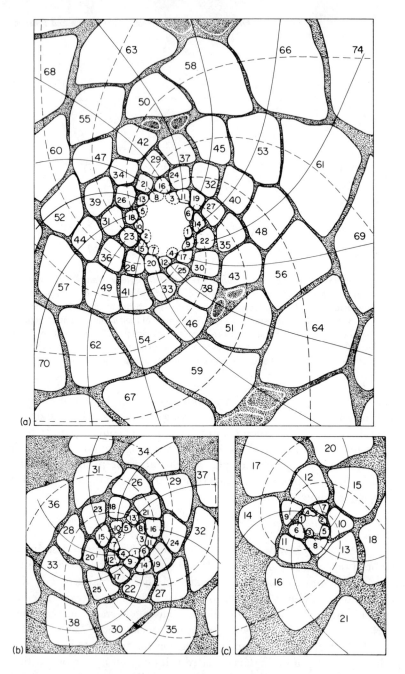

Fig. 7.6 Transverse sections, drawn at the same magnification, of the apical regions of branches of different sizes in *Araucaria excelsa* representing different orders of branching. These show different parastichy patterns (a) 8+13, (b) 5+8, (c) 3+5, without significant differences in average angular divergence. (Adapted from A. H. Church. 1904. On the Relation of Phyllotaxis to Mechanical Laws. Williams and Norgate, London.)

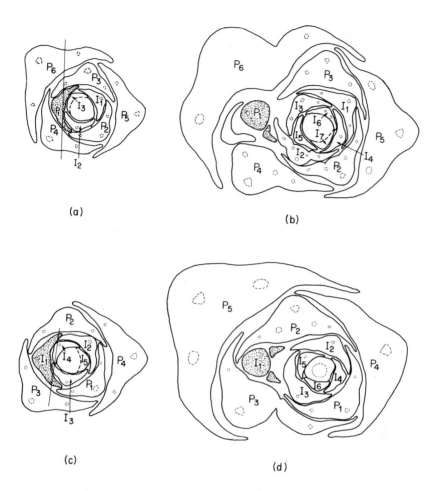

Fig. 7.7 The effect of isolating a leaf primordium of *Lupinus albus* (lupin) on the positions of subsequent primordia. (a) Diagram of a normal shoot apex showing the position of the tangential incision (the line in the diagram) that would isolate the P_1 primordium from the meristem. (b) Transverse section of a shoot apex in which P_1 was isolated from the meristem by a tangential incision. Several younger leaf primordia were formed after the operation. Primordium I_1 occupies its correct position, but the next primordium, I_2, is abnormally close to the isolated P_1 position, and the angular divergence between I_1 and I_2 has been increased from the normal 137.5° to 163.5°. (c) Diagram of a normal shoot apex showing the position of the tangential incision (the line in the diagram) that would isolate the I_1 position from the meristem. (d) Transverse section of a shoot apex in which the I_1 position was isolated from the meristem by a tangential incision. Several new leaf primordia were formed after the operation. Primordium I_2 occupies its normal position, but I_3 is abnormally close to the isolated I_1 leaf, and the angular divergence between I_2 and I_3 has increased to 198°. This has resulted in a reversal of the direction of the genetic spiral, which up to this point had been acropetally clockwise. (a–c) ×26, (d) ×22. (M. and R. Snow. 1931. Phil. Trans. Roy. Soc., London, B, **221** : 1.)

close to P_1, or rather to the incision that separated P_1 from the apex. This was determined by a measurement of angular divergence, with increases in divergence of up to 29° being found. The phyllotactic pattern of *Lupinus* is 2 + 3, with the result that the contacts for I_2 are P_1 and P_2, whereas those for I_1 are P_2 and P_3. Thus it may be argued that P_1 has an influence only upon positions adjacent to it. On the other hand, it could be argued that the position of I_1 was established before the operation was carried out. To test this possibility, the Snows performed another experiment in which I_1 was isolated by a tangential incision (Fig. 7.7c, d). In this case, I_2, which had been shown not to be fixed, arose in its normal position, and I_3 arose in a position abnormally close to I_1 with an increased divergence angle from I_2 of up to 67°. Thus some of the divergence angles were in excess of 180°, and reverses in the direction of the phyllotactic helix were noted. These experiments established that preexisting leaf primordia influence the positions only of new primordia that arise in their immediate vicinity.

Wardlaw (1949a) has applied similar techniques to the study of phyllotaxy in the fern *Dryopteris* that have generally confirmed the observations of the Snows. These are of particular interest because in this plant the leaf primordia are widely spaced and could not possibly be influenced by physical contact. Incision of the I_1 position prevented the formation of a leaf at this site and thus the effect of the absence of I_1 upon the emergence of later leaves could be assessed (Fig. 7.8). The positions of I_2 and I_3 were unaffected by this operation; but I_4 arose out of its expected position and in proximity to the incision at the I_1 site. This result is actually an exact parallel of the Snows' because in *Dryopteris* the contacts of I_3 are P_1 and P_3 whereas those of I_4 are P_2 and I_1, the phyllotactic pattern being 3 plus 5. This experiment and others in which various combinations of leaf sites were incised supported the conclusion that the position at which a leaf primordium arises is influenced by preexisting primordia adjacent to it. Wardlaw further observed that the abnormally placed primordia often grew faster than those in normal positions and frequently outgrew older primordia. This suggested an inhibitory effect upon young primordia by older adjacent primordia. In order to investigate this relationship further, Wardlaw isolated various primordia laterally by pairs of radial incisions. In each instance the isolated primordium outgrew other primordia that were older and larger, thus enhancing the inhibition interpretation.

In point of fact, the experiments in which the emergence of new primordia was displaced towards the positions of preexisting primordia that had been surgically isolated or suppressed can be interpreted also in term of regions of inhibition surrounding developing primordia. Further, the limitation of leaf initiation to the margins of the apical dome

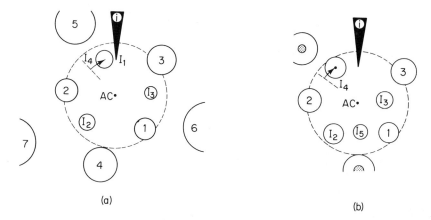

Fig. 7.8 The effect of destroying a presumptive leaf position in *Dryopteris dilatata* on the positions that subsequent leaves occupy. (a) A diagram of the shoot apex showing the predicted result of destroying the I_1 position by a knife cut (i). The leaves I_2 and I_3 are predicted to occupy their normal positions, but leaf I_4 should emerge closer to the I_1 operation site than in the normal apex. (b) An apex operated on as described above. It is seen that the leaves I_2 and I_3 occupy their predicted positions, as judged by their angular divergences and positions relative to their contact leaves, and that I_4 has emerged closer to the I_1 wound site. ×25. (Adapted from C. W. Wardlaw. 1949. Growth (suppl.) **13**: 93.)

is strongly suggestive of the operation of inhibition in the central region. This concept has been developed into a generalized field theory of phyllotaxy by Schoute (1913), and more recently by Richards (1948), and by Wardlaw. Wardlaw has elaborated this theory in the interpretation of the specific phyllotactic pattern of *Dryopteris*, and his discussion may be used in illustrating the application of the theory. According to Wardlaw, as each primordium is initiated, it is surrounded by a physiological field within which the inception of new primordia is inhibited. Although the fields are presumed to be chemical there is, as yet, no precise information on their nature. The fields are described as varying in intensity, decreasing with the age of the primordium, and each is presumed to have an intensity gradient decreasing from the center. A comparable field is seen as occupying the central dome of the shoot apex.

A simplified diagram in which the field theory is applied to a shoot apex is shown in Fig. 7.9; but such a diagram does not illustrate variations in field intensity nor does it reveal three-dimensional relationships. However, even without these important features, the diagram does show that areas available for new leaf formation must appear in a regular pattern as the shoot apex grows. The physiological field theory seems to apply particularly well to the apex of *Dryopteris* where it is necessary to account for the fact that leaf primordia do not occupy all of the space

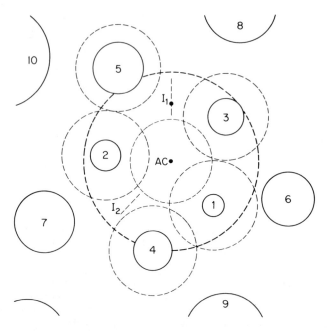

Fig. 7.9   A diagram of the shoot apex indicating the distribution of the hypothetical physiological fields that are associated with the terminal meristem of the shoot and the young emergent leaf primordia. The physiological fields are drawn as circles around the meristem and the leaves, and their lateral extent is arbitrarily fixed. The position of I_1 is the first to come to lie outside the existing fields as the shoot tip expands during growth, and I_2 is the next position to do so. (C. W. Wardlaw. 1949. Growth (suppl.) **13**: 93.)

that would appear to be available to them. The principle, however, applies equally well to apices in which the leaf primordia are more closely packed.

The results obtained by the Snows in *Lupinus* could be given the same interpretation as have those from *Dryopteris;* but the Snows themselves have preferred a different explanation. They regard the major factor limiting primordium initiation as space, and consider that a new leaf primordium arises in the first space that becomes available by reason of having attained sufficient width and distance from the summit of the apex. In this view, the experiments just described, in which the isolation of particular leaves or leaf sites influenced the positions of subsequent leaves adjacent to them, may be interpreted as having limited the lateral expansion of the isolated primordia and thus having altered the space relationships in the region of new leaf origin. Although this theory of the first available space appears to be an adequate vehicle for the description of leaf positioning in apices in which the leaves are closely packed, it is hardly satisfactory for those cases, such as *Dryopteris,* in which leaf primordia are widely spaced and availability of space could hardly be a limiting factor. Because the field hypothesis is equally applicable to both types of apices, and moreover includes a physiological rather than a geometrical mechanism for its operation, it does seem to be more acceptable than the available space theory, which has limited applicability.

Some years ago, a completely different kind of interpretation of the phenomena of phyllotaxy was advanced by the French botanist Plantefol (1948). This theory, which was widely accepted and elaborated upon by other workers in France, interprets phyllotactic patterns in terms of a number of foliar helices that wind around the stem in one direction and terminate in leaf generating centers in the apex (Fig. 7.10). This theory

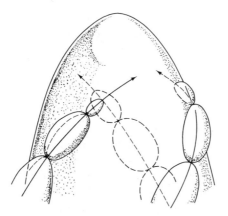

Fig. 7.10   A diagram of the shoot apex showing the hypothetical distribution of foliar helices. Three helices are shown with leaves being formed along each helix. (L. Plantefol. 1948. La Théorie des Hélices Foliares Multiples. Masson, Paris.)

of multiple foliar helices actually gave rise to the méristème d'attente concept of the shoot apex; the generative centers are thought to be located in the anneau initial. The stimulus to mitotic activity that initiates a leaf primordium is presumed to be transmitted acropetally along a foliar helix to the generative center in the meristem, but the activity of the several centers in an apex is held to be harmonized by an organizer located more apically in the meristem. In dicotyledons it is stated that there are generally two such foliar helices, one originating with each cotyledon, whereas in monocotyledons the number is variable but is usually greater than two. Because the continuity of the foliar helices is reflected in the apical bud in the contacts between adjacent leaves along the helix, contacts that may of course be removed by subsequent stem elongation, it is evident that the foliar helices have properties like those of contact parastichies long recognized by other workers apart from this theory. In fact one of the major criticisms of the theory of multiple foliar helices is that it is difficult to decide which of the contact parastichies in any shoot ought to be interpreted as foliar helices. Furthermore, many workers find it difficult to understand what a generative center means structurally if it is, in fact, propagated helically along the axis of the plant. Clearly what is involved is the sequential and disjunct occurrence of centers of mitotic activity associated with leaf initiation, and the fundamental point of this theory is that the location of these centers is causally related to a pattern in the already mature regions of the shoot. This is difficult to reconcile with clear evidence just cited that the placement of leaf primordia can be interpreted in terms of phenomena occurring within the shoot apex itself. The sum total of evidence would seem to suggest that the phyllotactic patterns described as foliar helices are better interpreted as parastichies and given a geometric rather than a functional or causal significance.

The idea that leaf sites may be determined by stimuli proceeding acropetally from older regions of the shoot is not restricted to those workers who support the theory of multiple foliar helices. After observing what appear to be partially differentiated leaf traces present in the stem before the leaves with which they will be associated have been initiated, a number of workers (Sterling, 1945; Gunckel and Wetmore, 1946) have suggested that such precocious traces might in fact transmit the leaf-initiating stimulus. A major difficulty with this hypothesis is that, although instances in which the leaf primordium and its trace develop essentially simultaneously are common, cases in which the trace clearly precedes the leaf have been described too infrequently to suggest that this could be a general mechanism. However, the presence of internal differentiation in the stem associated with presumptive leaf sites need not be inconsistent with the field theory if the physiological fields are envisioned as having depth as well as surface extension. The result of the operation of such a system would be the coordinated development of internal and external features regardless of the order of initiation.

A final question concerning phyllotaxy is the possible biological significance of the Fibonacci angle, which is characteristic of most species having a helical arrangement. If various phyllotactic patterns are plotted out using the divergence angle of $137.5°$, it will be seen that each primordium is placed between its contacts such that there is a constant ratio between angular distance from the primordium to the older and to the younger of the contacts. This system can continue indefinitely without any change in this ratio. This suggests the possibility that the ratio of angular distances in some way reflects the relative intensities of the fields of the contact leaves and the decrease in intensity as the primordium ages. The fact that in actual apices the angle fluctuates widely, with only the average approximating the ideal, supports the contention that biological factors such as this and not purely geometric considerations underlie the widespread occurrence of this angle.

General comment

The field theory has been presented in this chapter as the most plausible hypothesis to explain the placement of leaf primordia, at least in species with helical phyllotaxy. It may also have broader implications. In Chapter Six it was shown that the terminal meristem of the shoot functions as an integrated unit in spite of the fact that any portion of it, even a very small one, is capable of forming a new apex if suitably isolated from surrounding tissues. The operation of a field of inhibition, possibly generated by the central cells of the apex, could provide a mechanism by which the meristem is integrated; that is, kept free from outgrowths. If there is an intensity gradient from center to margins in this central field, outgrowths might be anticipated at the periphery of the meristem where they do indeed occur. Similar inhibitory fields around the peripheral outgrowths, operating in conjunction with the central field could explain why the primordia arise in a regular pattern and not haphazardly.

The fields are probably most easily visualized in terms of the production of inhibitory substances; but the nature of such substances has not been explored, nor is there clear proof that they actually occur. The fields might, in fact, result from the withdrawal or diversion of nutrients or essential growth factors. Moreover, as pointed out earlier, these two alternatives may not be mutually exclusive. The next logical step in the development of this important concept is the direct demonstration of the existence of fields and the elucidation of their biochemical basis. This step is long overdue; but it is extremely difficult because of the small size of the apical region and the probable low concentration of substances involved. In this connection, it is possible that careful ultrastructural studies could could reveal important facts about the metabolic states of particular groups of cells in the apex and the distribution of plasmodesmata through which complex metabolites might move in preferred pathways.

The field theory offers a working interpretation for the regulation of outgrowths from the shoot meristem and their positioning. There remains the further problem of how such outgrowths can acquire highly distinctive characteristics which convert them into determinate and dorsiventral leaves rather than additional shoot apices. The problem is further complicated by the fact that the same apex often produces other appendages which do form shoot apices, that is, buds. It is to this question that the next chapter is devoted.

REFERENCES

Church, A. H. 1904. On the relation of phyllotaxis to mechanical laws. Williams and Norgate, London.

Gunckel, J. E., and R. H. Wetmore. 1946. Studies of development in long shoots and short shoots of *Ginkgo biloba* L. II. Phyllotaxis and the organization of the primary vascular system; primary phloem and primary xylem. Am. J. Botany **33**:532–543.

Plantefol, L. 1948. La théorie des hélices foliares multiples. Masson et Cie, Paris.

Richards, F. J. 1948. The geometry of phyllotaxis and its origin. Symp. Soc. Exptl. Biol. **2**:217–245.

———. 1951. Phyllotaxis: its quantitative expression and relation to growth in the apex. Phil. Trans. Roy. Soc. London Ser. **B235**:509–564.

Schoute, J. C. 1913. Beiträge zur Blattstellungslehre. Réc. Trav. Bot. Néerl. **10**:153–325.

Schüepp, O. 1916. Untersuchungen über Wachstum und Formwechsel von Vegetationspunkten. Jahrb. wiss. Botanik. **57**:17–79.

Snow, M., and R. Snow. 1931. Experiments on phyllotaxis. I. The effect of isolating a primordium. Phil. Trans. Roy. Soc. London Ser. **B221**:1–43.

Steeves, T. A., and W. R. Briggs. 1958. Morphogenetic studies on *Osmunda cinnamomea* L.—The origin and early development of vegetative fronds. Phytomorphology **8**:60–72.

Sterling, C. 1945. Growth and vascular development in the shoot apex of *Sequoia sempervirens* (Lamb.) Endl. II. Vascular development in relation to phyllotaxis. Am. J. Botany **32**:380–386.

Sussex, I. M. 1955. Morphogenesis in *Solanum tuberosum* L.: apical structure and developmental pattern of the juvenile shoot. Phytomorphology **5**:253–273.

Wardlaw, C. W. 1949a. Experiments on organogenesis in ferns. Growth (suppl.) **13**:93–131.

———. 1949b. Further experimental observations on the shoot apex of *Dryopteris aristata* Druce. Phil. Trans. Roy. Soc. London Ser. **B233**:415–451.

EIGHT

Organogenesis in

the Shoot—The Determination

of Leaves and Branches

Having considered some of the factors involved in the positioning of successive leaves at the shoot apex, we must now turn our attention to the development of the individual leaf as a distinctive organ primordium. In particular, as pointed out earlier, it is necessary to investigate the nature of the influences that act upon outgrowths from the meristem, causing them to develop as leaves, and the nature of the response within the outgrowth, which results in the characteristic pattern of leaf development. Several experimental techniques have been brought to bear upon these questions, and, though no final answers have been achieved, a considerable body of relevant information has been amassed. At the same time a further problem must be explored. In addition to the regular formation of leaves, it is characteristic of the shoots of all but a few vascular plants to give rise to a succession of branches so that the whole shoot becomes a ramifying system. Previously the question was raised, and left unanswered: "Why does the shoot meristem form appendages or outgrowths at all?" To this must be added the further question: "If outgrowths are initiated, why are they not all alike?" Clearly the difference between a determinate and dorsiventral leaf and a branch that is a replica of the main axis is a striking one and it is important to seek an explanation for this difference in the initiation or early development of both types of appendages. This chapter is devoted to a consideration of these questions.

Leaf determination

In many ways, one of the most revealing approaches to the study of leaf development is that in which the partially developed organ is removed from the plant and allowed to continue its development on a culture medium of known composition in complete isolation from the parent organism. When this experiment was carried out on *Osmunda cinnamomea*, the rather surprising result was that immature leaves, including those as little as one millimeter in length, produced small mature leaves on a culture medium containing mineral salts and sugars but no complex organic supplement (Steeves and Sussex, 1957) (Fig. 8.1).

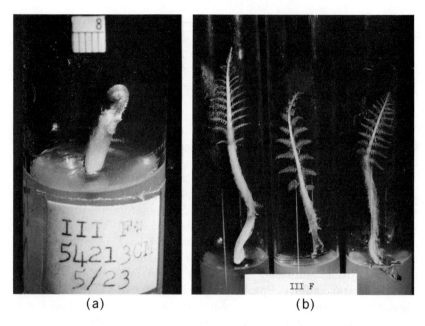

Fig. 8.1   *Osmunda* leaf cultures. (a) An excised leaf primordium growing on an agar culture medium. The leaf is just starting to uncoil. (b) Examples of leaves which have developed to maturity in sterile culture. (a) ×2, (b) ×0.6. (T. A. Steeves and I. M. Sussex. 1957. Amer. J. Bot. **44**: 665.)

Subsequently similar results were obtained with a number of other species of ferns and with several species of flowering plants (Fig. 8.2). This experiment clearly demonstrates that early in its development the leaf becomes essentially self-controlling or autonomous, and, in the presence of a supply of simple nutrients, is able to complete its characteristic pattern of development. The resulting mature leaves are ordinarily much smaller than those that mature on the intact plant and their morphological complexity may be reduced, but they are unquestionably determinate and dorsiventral leaves.

Fig. 8.2 Angiosperm leaf culture. An excised leaf of *Helianthus annuus* (sunflower) that has developed to maturity in sterile culture. ×4. (T. A. Steeves et al. 1957. Science **126**: 350.)

When, however, an attempt was made with *Osmunda* to discover whether the leaf primordium is autonomous from the time of its inception, a rather different result was obtained. If the youngest ten primordia in the shoot apex (P_1 to P_{10}) are explanted to culture medium, many give rise to shoots and ultimately to whole plants, after the initiation of roots, rather than to leaves (Steeves, 1961) (Table 8.1) (Fig. 8.3). The proportion of primordia that give rise to shoots is very high in the case of P_1 and P_2 and diminishes progressively in older primordia. P_{10} invariably develops as a leaf. Thus it may be concluded that the ability to develop as a leaf without further control from the rest of the plant is acquired by the primordium early in its development but not necessarily at its inception. In two species of angiosperms that have been investigated (*Helianthus*

TABLE 8.1

FATES OF EXCISED PRIMORDIA OF *Osmunda* CULTURED ON AGAR MEDIUM CONTAINING
MINERAL SALTS AND 2 PER CENT SUCROSE. (RESULTS FROM ONE EXPERIMENT.) (FROM
STEEVES, 1966.)

Primordium	Leaves	Shoots*	Doubtful or no growth
P_1	2	7	11
P_2	2	12	6
P_3	4	10	6
P_4	4	11	5
P_5	8	11	1
P_6	12	8	0
P_7	16	4	0
P_8	17	1	1
P_9	19	0	1
P_{10}	20	0	0

*In many cases whole plants.

annuus and *Nicotiana tabacum*) the youngest viable primordium explanted
(P_2) has consistently developed as a leaf, indicating that, at least in these
species, primordial destiny is established at a very early stage.

It is important to recognize the significance of the change effected
in these minute primordial outgrowths as they acquire autonomy. Initially, in the ferns at least, the primordium does not differ in its potentialities
from the shoot meristem that produced it. Quickly, however, it has imposed upon it a different and more specialized pattern of development.
Moreover, what is imposed upon it at this stage is not a particular set of
morphological characteristics but a program for future development
that it is then able to carry out by itself to the formation of a mature
organ. This process whereby a group of totipotent meristematic cells
becomes established in a particular developmental pathway is often
called *determination* in conformity with the usage of this term for similar
phenomena in animal development. In fact the determination of a leaf,
which arises in a definite site and has a limited period of development
followed by complete maturation, has many similarities to the determination of organ primordia in animals. Although in the angiosperm
species studied, determination seems to occur at an earlier stage than in
ferns, there is no reason to expect that the phenomenon is fundamentally
different in the two groups.

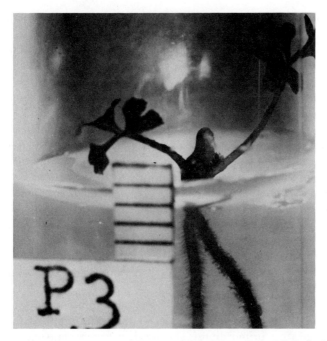

Fig. 8.3 Transformation of an excised leaf primordium into a shoot. A P_3 leaf primordium excised from the bud of *Osmunda cinnamomea* and grown in sterile culture has developed into a shoot on which several leaves have been initiated and from which roots have developed. ×5. (T. A. Steeves. 1961. Phytomorphology **11** : 346.)

Surgical investigations

Some useful insights into the process of leaf determination have also been obtained by the use of surgical techniques in which, because the primordium is only partially isolated by incisions, it is possible to deduce something of the source of determining influences. In the fern *Dryopteris*, Wardlaw (1949) isolated the I_1 position from the center of the meristem by a wide and deep tangential incision. The I_1 position thus isolated in most cases gave rise to a shoot bud rather than to a leaf, indicating that it was not at this stage determined. Later, in the same fern, Cutter (1956) was able to cause the three youngest visible leaf primordia (P_1 to P_3) to develop as shoots by isolating them as Wardlaw had done with the I_1 position or by isolating them from all lateral contacts on a plug of tissue by four deep, vertical incisions. Here, as in the *Osmunda* culture studies, determination of the primordium as a leaf must occur sometime after its emergence.

In the ferns that have been investigated the leaf primordium undergoes a certain amount of characteristic development prior to determination. This fact might suggest that it is determined only after it has achieved a particular level of morphological organization or that such organization is what constitutes determination. However, this possibility has been ruled out, at least in the case of *Osmunda*, by the demonstration that pieces of leaf primordia in many cases give rise to entire leaves when grown on nutrient culture medium (Steeves, 1966). The leaf fragments were obtained by subdividing the primordia P_4 to P_{10} in various ways so that none contained the total structural configuration of the intact primordium. In a few cases three leaves have been obtained in this way from a single primordium, from the apex and two halves of the base, respectively. It begins to appear that the phenomenon of determination may be expressed in the primordium at the cellular level rather than at higher levels of organization, although this is certainly an extrapolation from existing data.

In considering the phenomenon of leaf determination, it also is necessary to seek the possible sources of the stimulus that produces this important change in a primordium. The surgical experiments cited, because they involved principally the isolation of the primordium or primordium site from the central region of the shoot apex, could be interpreted as indicating that the most distal cells of the shoot meristem are the major source of the determining stimulus. Indeed, they have usually been so interpreted. Recently this interpretation has been supported by Hicks and Steeves (1969) using shoot apices of *Osmunda* growing on nutrient culture. When all leaf primordia were removed from the apex, a new primordium quickly arose in the I_1 position, and within two weeks this primordium had been determined as a leaf that could be shown to be autonomous by explanting it separately. Thus determination had occurred in the absence of any older leaf primordia. If the I_1 position was isolated from the center of the apex by a deep tangential incision into which a chip of mica was inserted, in 75 percent of the cases a shoot arose instead of a leaf. There is thus clear evidence of a major influence emanating from the distal cells of the apex.

Evidence pertaining to the nature of the stimulus, or at least to the mode of its transmission, has also been obtained in this series of experiments. If the isolating incision was made by a very fine knife with a drop of culture medium covering its surface to prevent desiccation and no mica chip was inserted, approximately 80 percent of the resulting outgrowths were leaves. Histological examination revealed that in most cases a grafting process had established a tissue continuity across the incision in the inner tissues of the apex, but not in the surface layer of large cells, and that this grafting process began within the first week after the incision. In the cases examined, there was a correlation between the failure of grafting to occur and the development of a shoot rather than a leaf. The influence that proceeds from the distal cells of the apex thus appears to require tissue continuity for its transmission and its effective transmission through inner tissues of the apex where the surface layers have been severed supports the conclusion of Wardlaw and Cutter (1956) that a deep incision is much more effective in isolating a primordium that is a shallow one alone.

The fact remains, however, that in these experiments some 25 percent of the I_1 positions isolated from the center of the meristem by a non-grafting incision formed leaves even though it is known that this position is not determined as a leaf at the time of this incision. This suggests that other factors, or factors from other sources, may at least participate in the determination process, and there is other experimental evidence to support such a view. One example may be cited to illustrate this point. In a study of induced apogamy in the fern *Pteridium*, Whittier (1962) observed that among the sporophytic structures that developed on haploid gametophytes without fertilization, there were a number of individual leaves. Careful examination of these structures and of their development failed to reveal the presence of a shoot apex at any stage of their ontogeny. Thus it is possible for a determinate and dorsiventral leaf to arise in a growing system without the participation of a shoot apex; and it is therefore difficult to visualize determination as a process that depends exclusively upon the meristematic cells of the apex. This point of view has more recently received considerable emphasis in experiments performed by Kuehnert (1967) in which third primordia (P_3) of *Osmunda*, which when explanted to nutrient culture normally give rise to leaves in only about 25 percent of the cases, produced leaves twice as frequently when cultured in contact with older and clearly determined primordia, notably P_{10} to P_{12}. It is presumed that some influence, probably chemical, passes from the older primordium to the younger where it exerts a determining influence, or at least a stabilizing influence upon determining processes already underway. In any event the exclusive role of the shoot meristem in determining leaf primordia is challenged by these observations.

The sum total of information pertaining to the origin and nature of the determining stimulus does not lend itself to a simple or clear-cut interpretation. Rather, it begins to appear that multiple factors may be involved and that in the intact shoot apex determination is the net result of a number of influences to which the primordium is subjected. Undoubtedly the shoot meristem itself plays a prominent role, but in its absence other stimuli may be adequate to bring about determination. In a system such as this it is evident that experimental results must be interpreted with extreme caution.

Dorsiventrality

Perhaps the most striking structural feature of most, although not all, leaves is their bilateral symmetry or dorsiventrality. Developmentally the appearance of this feature is of great interest because the initially radial outgrowth of the leaf becomes flattened in a plane that is tangential to the circumference of the shoot meristem, suggesting that the one-sided relationship of the primordium to the meristem may be important in initiating this characteristic. Indeed, Wardlaw (1949) has suggested that an inhibitory influence from the meristem may retard growth on the adaxial side of the primordium as compared with the abaxial side, and that this growth inequality may be fundamental in the initiation of dorsiventrality. Experiments which he and later Cutter carried out on *Dryopteris*, some of which have been described, are consistent with this view. In *Solanum tuberosum* (potato) Sussex (1955) was able to provide additional confirmation of this hypothesis when, after isolation of the I_1 position from the remainder of the meristem, he observed the development of organs that were determinate in growth but radially symmetrical, that is, centric leaves (Fig. 8.4). Centric leaves have also been obtained in *Epilobium* by the Snows (1959) and in *Sesamum* following a similar operation. In *Sesamum*, Hanawa (1961) obtained both a dorsiventral leaf and a centric leaf from the same primordium by incising a P_1 tangentially across its apex. The dorsiventral leaf arose from the half primordium in continuity with the apex and the centric leaf developed from the half on the abaxial side of the incision.

However, there is evidence supporting another interpretation of the origin of bilateral symmetry in leaves. The experiments of Whittier that were discussed earlier, in which perfectly dorsiventral leaves arose as apogamous outgrowths from fern prothalli, must be interpreted differently because no shoot meristem was present. Similarly Steeves (1962) has reported that excised leaf primordia in culture sometimes undergo a period of development during which their symmetry is radial following which they become dorsiventral, again in the absence of a shoot meristem (Fig. 8.5). In these cases the bilateral symmetry of the organ has been interpreted as a consequence of its determination as a leaf, that is, an inherent part of the leaf pattern of development acquired in the process of determination, rather than as a special feature requiring its own induction. But in the experiments described on potato apices, the development of determinate but centric leaves led to the conclusion that the primordia had been determined as organs of limited growth but had not

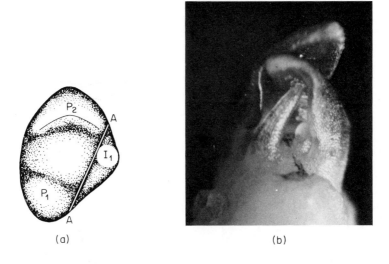

(a)   (b)

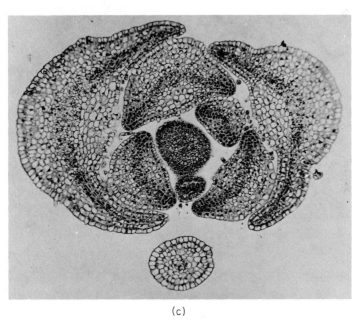

(c)

Fig. 8.4 Development of a leaf of *Solanum tuberosum* (potato) as a radially symmetrical organ. (a) Diagram of the shoot apex showing the incision that separated the site where the next leaf (I_1) would emerge from the terminal meristem and the two older leaf primordia P_1 and P_2. The radial shape of the I_1 leaf that will develop in the isolated site is indicated in the diagram. (b) External view of a shoot apex operated on as in (a) with a radially symmetrical I_1 primordium at the front left, and the larger dorsiventral leaves P_1 and P_2 behind. (c) Transverse section through the shoot tip shown in (b). The radial organ is at the bottom of the illustration, P_2 is to the left, and P_1 is on the right. (b) ×28, (c) ×75. (I. M. Sussex. 1955. Phytomorphology **5**: 286.)

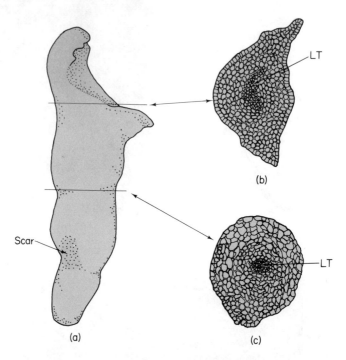

Fig. 8.5   A leaf primordium of *Osmunda cinnamomea* in which dorsiventrality developed after excision from the plant. (a) External view of the excised primordium after growth in sterile culture. The primordium was excised at the P_3 stage. The adaxial surface of the upper part of the leaf faces 90° away from the original adaxial face which is indicated by the scar on the base. A root is growing downward from the base near the scar. (b) Transverse section of the upper dorsiventral part. (c) Transverse section of the lower radial part of the leaf. The levels of the sections (b) and (c) are shown in (a). Key: LT, leaf trace. (a) ×25, (b, c) ×40. ((a) drawn from T. A. Steeves. 1962. Phytomorphology **11** : 364; (b, c) drawn from T. A. Steeves. 1962. Regeneration. D. R. Rudnick [ed.] Ronald Press, New York.)

undergone the induction of dorsiventrality at the time of the operation. In this second view, the total determination of a leaf could be looked upon as consisting of more than one step, with various features of the leaf pattern being induced sequentially and perhaps independently. These alternative interpretations, based as they are upon separate experiments in different organisms, suggest that the process of leaf determination will not be fully understood until much further work has been done.

Branching

In considering the initiation of appendages that become shoots —that is, branches—it must be kept in mind that there appear to be more than one type of branching in the vascular plants. In fact two apparently distinct types may be identified. In the Psilopsida (*Psilotum* and *Tmesipteris*), Lycopsida (*Lycopodium* and *Selaginella*), and many ferns, branching commonly occurs when the shoot meristem becomes separated, equally or unequally, into two growing centers, each of which develops a shoot. Thus if the division of the apex is equal and the subsequent growth of the two branches equivalent, an equal forking of the shoot results (Fig. 8.6a). On the other hand, if the subdivisions of the

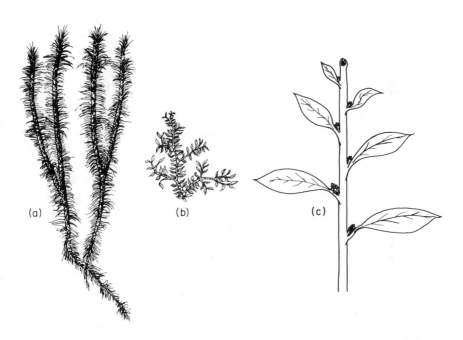

Fig. 8.6   Branching patterns in the stem of vascular plants. (a) Equal branchings of the terminal meristem in *Lycopodium lucidulum*. (b) Unequal terminal branching in *Selaginella kraussiana*. (c) Axillary branching in an angiosperm.

apex are unequal, or subsequently develop unequally, smaller and larger shoots result from each branching, and in many cases the succession of larger branches may form a monopodial axis with apparent side branches (Fig. 8.6b). In many ferns, often in accompaniment with terminal branching, and in the seed plants, lateral axes arise from buds situated in or near the axils of leaves, or, in ferns, often in interfoliar positions (Fig. 8.6c).

Terminal branching

Terminal branching has been described as resulting from the separation of the shoot meristem into two equal or unequal portions, each of which becomes the apex of a branch axis. Although this is easy enough to visualize descriptively, it is quite another matter to understand how this kind of subdivision can occur if the meristem is a single, integrated growth center, as experimental studies reviewed earlier have indicated. Some of these experiments in fact have been revealing as to the mechanism of terminal branching. In many vascular plants, if the shoot meristem is surgically subdivided, the various portions reorganize and form complete meristems that initiate branch shoots. In several ferns it has been shown that puncturing of the apical cell, together with the attendant necrosis of a group of immediate derivatives, will produce the same result. Thus if the integration of the shoot meristem is artificially broken down, a kind of terminal branching results; but there is nothing to suggest that any injury or related phenomenon is involved in the natural occurrence of terminal branching.

There is very little information concerning the structural changes that take place in terminal branching and the evidence suggests that there may be more than one type of process involved. In *Osmunda* where terminal branching is equal but infrequent, early stages of branching show two active growth centers on the flanks of the meristem, whereas in the center the original apical cell and its immediate derivatives can still be found, but subdivided into smaller cells in the same way that peripheral cells of the prismatic layer normally become subdivided. A similar process has been described in *Selaginella willdenovii* (Cusick, 1953) where, instead of a single apical cell, there is a band of enlarged initial cells across the summit of the apex. Branching occurs when several of these cells near the center of the band begin to differentiate, leaving two unequal growth centers that form branch shoots. This method of terminal branching would be equivalent to that induced by surgery if it could be shown that the differentiation of the most distal cells precedes the outgrowth of lateral flanks and presumably causes it, or at least permits it, to occur. However, although no precise information is at hand, it seems more likely that the establishment of the lateral growth centers occurs first and that their combined influence promotes the differentiation of the distal cells. Such a mechanism would be consistent with the observation in some lower plants, including both *Osmunda* and *S. willdenovii*, that there is a progressive increase in apical size as the plant develops that is limited by the occurrence of successive branchings. It might be postulated that the meristem remains as a single unit in its organization up to some maximum size, at which point two new organizational centers replace the original.

In some other plants that branch terminally, the apical cell of the original meristem is described as continuing, and one of its relatively recent derivatives becomes the initial of a smaller branch. In the fern *Pteridium* (Gottlieb and Steeves, 1961), this type of branch formation has been observed in the initiation of short lateral branches on the main rhizome, whereas in the equal branching of the main axis the former method has been described. The mechanism by which a second apical cell can become established and initiate a new apex while still in proximity to the original apical cell is difficult to visualize; and furthermore it is not apparent why such an outgrowth should not develop as a leaf. From the foregoing it is evident that the phenomenon of terminal branching has many interesting aspects, particularly concerning the nature of apical integration and how it is maintained or broken down in an orderly way; but until more than a few isolated cases have been described in detail, it is pointless to attempt any generalizations.

Lateral branching

For those plants that do not branch terminally, but rather initiate branches from buds that appear to develop at some distance from the shoot meristem, the histological phenomena have been much more fully investigated. Wardlaw (1943) has shown that in several ferns lateral shoots, occupying what have been called interfoliar positions, arise from groups of undifferentiated cells, *detached meristems*, that did not undergo development while leaf primordia were emerging around them, but also did not differentiate (Fig. 8.7). These detached meristems have a developmental continuity with the apical meristem of the shoot of which they are in fact portions that are somehow inhibited in the region of the apex and may even remain inhibited in the mature shoot. In some cases they never develop unless the main apex is removed, in which case they give rise to small lateral shoot apices, or they may develop in this way at some distance from the main apex without its removal. Lateral branches in seed plants typically arise in axillary positions; and in the majority of those investigated the origin has been found to be remarkably similar to that described by Wardlaw in ferns (Fig. 8.8). Garrison (1955) has described axillary bud origin in a number of dicotyledons of diverse taxonomic affinities. As successive leaf primordia are formed at the shoot apex, it may be seen that the cells in their axils do not undergo cell enlargement and vacuolation like the cells around them, but remain as detached meristems which, when removed by further development

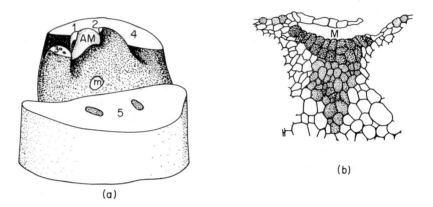

(a)

(b)

Fig. 8.7   The shoot apex of the fern *Matteucia struthiopteris* showing the distribution of meristematic tissue. (a) The apical meristem which extends over the summit of the shoot reaches into the axils of the four youngest leaves. Above the axil of leaf 5 cells have begun to differentiate but a portion of the meristem has persisted as a detached meristem. Only the bases of leaves 3–5 are shown in the figure. (b) A section of a detached meristem. Key: AM, apical meristem, M, detached meristem. (a) ×20, (b) ×85. (C. W. Wardlaw. 1943. Ann. Bot. **7**: 171, (a) modified.)

Fig. 8.8   Stages in the differentiation and development of detached meristems of *Solanum tuberosum* (potato). (a) Origin of the detached meristem. Divisions in surrounding cells delimit a small group of meristem cells above the axil of a leaf primordium on the right side of the shoot apex. (b) An early stage in the growth of the detached meristem as an axillary bud. ×225. (I. M. Sussex. 1955. Phytomorphology **5**: 253.)

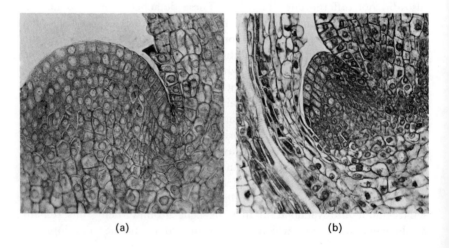

(a)   (b)

from the original apical meristem, enlarge and become organized into shoot apices with leaf primordia. Commonly at this time the dominance of the main apex is expressed in the inhibition of further development of the lateral apices and they remain as axillary buds, often for long periods and sometimes permanently, unless the main apex is removed.

In certain respects the origin of lateral branches from detached meristems that have a developmental continuity with the shoot meristem of the original axis resembles unequal terminal branching in that a portion of the meristem initiates a small branch while the major part of the meristem continues its development. The main difference is one of timing, in that the detached meristem is restrained from organizing a shoot apex until further growth has left it outside the region of the shoot apex. There is also the further difference that detached meristems show a more or less constant relationship to leaf primordia, a relationship that is lacking in terminal branching. It has been suggested that the detached meristem pattern may be derived phylogenetically from that of unequal terminal branching. On the other hand there are some reports (e.g., Majumdar, 1942) of a different method of initiation of lateral buds in seed plants. In these it is stated that axillary buds do not arise directly from the shoot apex as detached meristems but rather are initiated at some distance from the apex from cells in the leaf axil that are already partially or completely differentiated. These reports deserve to be more fully investigated and an attempt should be made to determine whether this pattern has a widespread distribution. However, in view of the fact that adventitious buds are well known to arise from a variety of mature tissues in stem, leaf, and root as well as from wound callus, it is not startling to contemplate such an origin from differentiated cells for axillary buds. Only their regular occurrence and constant placement in relation to leaves would pose problems of interpretation.

Several intriguing problems concerning the origin and development of buds arise because of the interesting relationships between buds and leaves. In the first place, there is the question of why there should be two types of appendages, that is, if the conditions in the shoot apex are such as to determine some appendages as leaves, why do they not determine all appendages in this way? In addition, the nearly constant spatial relationship to leaves suggests that there may be functional interactions between the two. These problems have been explored to some extent by experimental methods. In experiments discussed previously, Wardlaw and Cutter have shown that incipient leaf positions and even emergent leaf primordia in certain ferns are capable of initiating buds if they are suitably isolated from their surroundings, particularly from the influence of the shoot meristem itself. In other experiments in which the phyllotactic sequence has been altered by surgery, Wardlaw (1949) has shown that leaves can be developed in locations that may be identified as the normal sites of detached meristems, and hence buds. Thus any part of the shoot meristem seems to be able to initiate either a bud or a leaf.

In the majority of cases investigated, buds do not grow out from the intact meristem like leaves, but rather are delayed and emerge in the subapical region. Thus they are not exposed to the factors that determine other outgrowths that develop immediately as leaves. It is at least possible that the different morphology of leaves and buds is a result of this difference in timing of development and that the key to understanding this difference lies in the mechanism that inhibits the development and the differentiation of certain areas of the meristem, the detached meristems, until they are out of the apical region. The close association of buds and leaves might suggest that leaf primordia may be involved in this delay. But because in *Dryopteris* Wardlaw has shown that buds will grow out in the region of the meristem if the shoot apical cell is punctured while the adjacent leaves remain intact, the postulated role of leaf primordia must be more involved with the delay in the differentiation of the detached meristem than with the inhibition of its growth.

Other experiments in the angiosperms that bear upon the relationship between bud and leaf may now be considered. Various experimental procedures have been shown to alter the positions at which leaf primordia arise, that is, to change the phyllotactic pattern. It has been noted repeatedly in such cases that the axillary bud is associated with the leaf primordium in its unaccustomed position, indicating that the normal association is not one of chance. Because in many cases the leaf primordium distinctly precedes the bud in origin, the leaf must be considered the primary component in this association. In *Epilobium* and several species of *Salvia* the Snows (1942) have shown that if a young leaf primordium (P_1) is suppressed by incising it so that it fails to develop, its associated lateral bud, although not damaged by the operation, fails to appear. If, however, the leaf primordium is merely cut off at about the level of the meristem, the remaining base is often adequate to promote the formation of the bud. Similarly, if the leaf is isolated from the center of the meristem by a tangential incision immediately adjacent to the primordium so that the bud position is left intact, the bud likewise fails to develop. The Snows have interpreted these experiments as demonstrating the important role of the leaf in determining its own axillary bud. Although the experiments do indicate that the leaf is involved, they could equally well be interpreted as revealing the role of the leaf in preventing the differentiation of the potential detached meristem. Wardlaw has suggested the distribution of tensile stress resulting from the development of leaf primordia as a possible mechanism for their action in this respect. This interesting proposal certainly deserves further investigation in a variety of species, along with other possible mechanisms of bud determination.

There is, however, evidence that the regulation of bud initiation is, or at least can be, more complex than is indicated from the experiments thus far reported. This evidence is derived from a study of the distribution of bud primordia in the apices of certain angiosperms, a few examples of which will serve to illustrate the point (Cutter, 1966). In some plants buds are initiated in the axils of certain leaves only, and often there is a regular pattern of their distribution. Thus whatever the factors that determine the initiation of a bud, they cannot be such as to be present in every leaf axil. There are other instances, as in *Nymphaea*, in which buds arise in place of leaf primordia in the phyllotactic sequence so that leaf-determining and bud-determining stimuli must be operative in close proximity in the same apex. In *Hydrocharis* conspicuous axillary buds are formed on the intact terminal meristem. These examples, and others like them, need not indicate complications in bud determination that are applicable to all plants, that is, to the basic mechanism of induction; but they certainly must be explained before any sweeping generalizations are offered.

General comment

The phenomenon of leaf determination is one which invites a rigorous analytical investigation, particularly since the leaf primordium is an easily isolated developmental unit. The work completed to date has done little more than frame the questions which need to be answered. The evidence presently available does not suggest that a single, highly specific stimulus is responsible for bringing about determination; yet, in different experiments, both the shoot meristem and older leaf primordia have been shown to exert an influence, probably by chemical means. Meanwhile, it would be unwise to overlook the possible importance of the overall environment of the young primordium in inducing the leaf pattern. What, for example, is the consequence of the fact that the primordium is a rapidly growing center surrounded by areas which are inhibited from growing out in the same way? Any surgical interruption of such a system would disrupt this environment and could lead to erroneous conclusions about specific stimuli and their source. What is urgently needed for further progress is precise information about the detailed biochemistry of the shoot apex. What substances are produced in localized regions and how are they transmitted to other areas? Highly refined techniques will be required to obtain this information; but the results would certainly justify the effort expended in perfecting them.

At the same time it is equally important to know precisely what happens to the leaf primordium when it is determined. Clearly changes occur which do not appear to be structural and which quickly become independent of the inducing stimuli. Indeed, it is a pattern of development which is induced rather than a particular structural feature. Again the fact that primordia can be isolated easily, and even allowed to develop in isolation, should facilitate this investigation, however exacting it may be technically. An important question here is the extent to which the leaf primordium is determined, that is, how much of its potential morphological complexity is established by the determination. The subsequent chapter will analyze this problem in some detail.

REFERENCES

Cusick, F. 1953. Experimental and analytical studies of pteridophytes. XXII. Morphogenesis in *Selaginella willdenovii* Baker. 1. Preliminary morphological analysis. Ann. Botany (London) (N. S.) **17**:369–383.

Cutter, E. G. 1956. Experimental and analytical studies of pteridophytes. XXXIII. The experimental induction of buds from leaf primodia in *Dryopteris aristata* Druce. Ann. Botany (London) (N. S.) **20**:143–165.

———. 1966. Patterns of organogenesis in the shoot. *In* E. G. Cutter [ed.] Trends in Plant Morphogenesis, pp. 220–234. Longmans, London.

Garrison, R. 1955. Studies in the development of axillary buds. Am. J. Botany **42**:257–266.

Gottlieb, J. E., and T. A. Steeves. 1961. Development of the bracken fern, *Pteridium aquilinum* (L.) Kuhn. III. Ontogenetic changes in the shoot apex and in the pattern of differentiation. Phytomorphology **11**:230–242.

Hanawa, J. 1961. Experimental studies on leaf dorsiventrality in *Sesamum indicum* L. Bot. Mag. Tokyo **74**:303–309.

Hicks, G. S., and T. A. Steeves. 1969. In vitro morphogenesis in *Osmunda cinnamomea*. The role of the shoot apex in early leaf development. Can. J. Botany **47**:575–580.

Kuehnert, C. C. 1967. Developmental potentialities of leaf primordia of *Osmunda cinnamomea*. The influence of determined leaf primordia on undetermined leaf primordia. Can. J. Botany **45**:2109–2113.

Majumdar, G. P. 1942. The organization of the shoot in *Heracleum* in the light of development. Ann. Botany (London) (N. S.) **6**:49–81.

Snow, M., and R. Snow. 1942. The determination of axillary buds. New Phytol. **41**:13–22.

———. 1959. The dorsiventrality of leaf primordia. New Phytol. **58**:188–207.

Steeves, T. A. 1961. A study of the developmental potentialities of excised leaf primordia in sterile culture. Phytomorphology **11**:346–359.

——. 1962. Morphogenesis in isolated fern leaves. *In* D. Rudnick [ed.] Regeneration. 20th Growth Symp. pp. 117–151. Ronald, New York.

——. 1966. On the determination of leaf primordia in ferns. *In* E. G. Cutter [ed.] Trends in Plant Morphogenesis. pp. 200–219. Longmans, London.

Steeves, T. A., and I. M. Sussex. 1957. Studies on the development of excised leaves in sterile culture. Am. J. Botany **44**:665–673.

Sussex, I. M. 1955. Morphogenesis in *Solanum tuberosum* L.: Experimental investigation of leaf dorsiventrality and orientation in the juvenile shoot. Phytomorphology **5**:286–300.

Wardlaw, C. W. 1943. Experimental and analytical studies of pteridophytes. II. Experimental observations on the development of buds in *Onoclea sensibilis* and in species of *Dryopteris*. Ann. Botany (London) (N. S.) **7**:357–377.

——. 1949. Experiments on organogenesis in ferns. Growth (Suppl.) **9**:93–131.

Wardlaw, C. W., and E. G. Cutter. 1956. Experimental and analytical studies of pteridophytes. XXXI. The effect of shallow incisions on organogenesis in *Dryopteris aristata* Druce. Ann. Botany (London) (N. S.) **20**:39–56.

Whittier, D. P. 1962. The origin and development of apogamous structures in the gametophyte of *Pteridium* in sterile culture. Phytomorphology **12**:10–20.

NINE

Organogenesis in
the Shoot—Later Stages
of Leaf Development

The stages of leaf development discussed in the two previous chapters—the primordial stages, which culminate in a simple outgrowth at the margin of the shoot meristem, somewhat flattened on its adaxial face—give little indication of the diverse morphology of the mature leaves of various groups of vascular plants. The multipinnate frond of a fern, the needle leaves of many conifers, and the diverse simple and compound leaves of the angiosperms are remarkably similar in the period immediately following their inception. Thus the diversity of leaf form can be interpreted best through an understanding of the later, as opposed to primordial, stages of development. On the other hand the evidence here cited indicates that in the primordial stages the leaf undergoes a determination that confers upon it a considerable degree of autonomy in its later development. Does this imply that all of the diverse morphology of leaves must be thought of as originating in a process of determination at a relatively undifferentiated stage? The answer to this question is not an easy one, and the issues involved may best be exposed by examining some of the events of later leaf development and some of the experiments that have sought to interpret them.

The fern leaf

Description of development

There is a substantial body of information about later stages of leaf development in ferns, much of it collected from species that have also been used for experimental analysis, so that descriptive and experimental data may be correlated. One such example is the leaf of *Osmunda cinnamomea*, the early development of which has already been considered. In this fern the full developmental sequence of the leaf requires five growing seasons for its completion (Steeves, 1963) (Fig. 9.1). This is probably an extreme case, but undoubtedly it is not unique. Apical growth is long continued, even though it is limited, and is accomplished by a terminal meristem with a central apical initial. The growth of the apex is not uniform, however, and several phases may be recognized. An abaxial-adaxial disparity in growth, which reflects the initiation of dorsiventrality, causes an early curvature in the primordium that is retained at or just behind the leaf apex, giving it the form of a hook. This results from an inequality in cell number, at least in the ground tissue of the primordium, but this inequality is continuously equalized at the base of the hook so that a straight axis is produced. In the fourth growing season of a leaf, apical growth appears to be accelerated to such an extent that the equalizing process does not keep pace and the result is the rather rapid formation of a coiled axis or crozier (Fig. 9.1a). As the crozier is formed there is enlargement of the outer coils producing a loosening that allows space for the still-forming inner coils. This is accomplished by extensive cell division at considerable distance from the leaf apex with no net cell elongation. Surprisingly enough, this cell division maintains the imbalance in cell number set up just behind the apex so that the crozier does not uncoil at this time. As the leaf apex begins to form the crozier, it also starts the initiation of pinnae in acropetal sequence along each side of the axis. At the end of the fourth growing season, the apical meristem of the leaf ceases to function as such, the apical cell differentiates, and ultimately a terminal pinna is formed. Thus after a prolonged period of activity, the determinate character of the leaf apex is expressed.

The expansion of the leaf out of the apical bud and its final maturation occur in the fifth growing season immediately after a period of dormancy (Fig. 9.1b, c). In this phase, while cell division continues in the more apical coiled regions of the leaf, a wave of cell elongation begins to extend acropetally in the ground tissue of the leaf axis leading to enormous increases in length and to final maturation. As this process extends into the crozier, the coiled axis begins to straighten and the crozier uncoils.

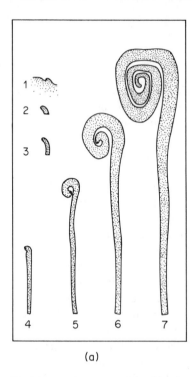

(a)

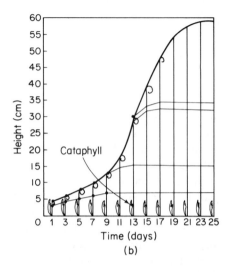

(b)

(c)

Fig. 9.1 Development of the frond of *Osmunda cinnamomea*. (a) Frond development from inception to the end of the fourth growing season: (1) soon after inception; (2) end of the first season; (3) end of the second season; (4) end of the third season; (5–7) developmental stages during the fourth season. The coiled crozier develops entirely during the fourth season. (b) Graphic record of the development of a vegetative frond of *Osmunda* during the fifth growing season. The cataphyll heights indicate the length of the original apical bud. The frond was marked at regular intervals, but only a few of the marks placed on the frond are plotted in this graph. Observe that the mark placed at the top of the crozier does not change position during the period of crozier elevation, and that the crozier begins to uncoil only between days 9 and 11. (c) An *Osmunda* plant showing fronds in various stages of crozier uncoiling. (T. A. Steeves. 1963. J. Linn. Soc., (Bot.) **58**: 401.)

Just before cell division ceases at any level in the crozier in advance of the wave of elongation, the long-perpetuated imbalance in cell number abaxially and adaxially is removed by the occurrence of extra divisions on the adaxial side. Thus when elongation is completed the rachis is straight with cell number and cell size equal on the two sides.

This brief account of the later phases of leaf development in *Osmunda* emphasizes the complexity as well as the long duration that may be found in the postprimordial stages. The precise timing of the sequential developmental events that are necessary for the successful completion of the leaf points to a rather elaborate controlling system and gives focus to the question of whether all of this pattern can be considered as a consequence of determination at the primordial stage. Until other species of ferns have been investigated in similar detail, it cannot be known how nearly typical the *Osmunda* pattern is; but what is known of other ferns suggests that the differences are largely ones of the duration and distinctness of phases. Certainly in some ferns the complete development of a leaf from initiation to maturity is accomplished in one growing season or less and, even in *Osmunda*, juvenile leaves of the sporeling develop very rapidly. The problem is to understand how the later phases of development are regulated; for this it is necessary to turn to experimental studies.

Experimental studies on Osmunda

In *Osmunda* some information on controlling mechanisms has been obtained by the use of sterile culture techniques in which leaves in various stages of development have been removed and allowed to continue in isolation. In this way, the capacity of the leaf to regulate its own development from the stage at which it is excised can be determined and its dependence upon extrinsic factors can also be evaluated. Caponetti and Steeves (1963, 1970) have investigated primordia excised at the end of their third growing season but prior to the beginning of crozier formation. The isolated development of these fronds is remarkably similar to that of natural fronds and it duplicates the details of crozier formation and loosening, pinna initiation, cessation of apical growth, uncoiling, and final expansion and maturation (Fig. 9.2). The resulting mature leaf, however, is only about one-tenth the normal height with one-third the normal number of pinnae, and development is completed in a fraction of the time taken by the normal frond. The reduced height results from reduction in cell number along the axis of the frond, because cell length measured in the ground tissue equals that of natural fronds.

Although the overall pattern of morphogenesis in cultured and natural fronds is remarkably similar, cellular processes show marked

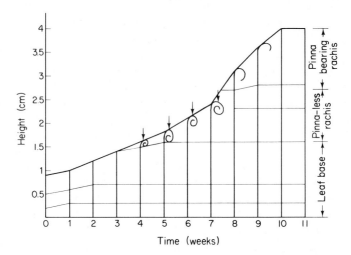

Fig. 9.2 Graphic record of the development of a frond of *Osmunda cinnamomea* excised from the plant at the end of the third growing season and grown in sterile culture to maturity. The crozier developed during the first three weeks of culture. It elevated in the next three weeks (observe the arrow indicating the position of a mark placed on the top of the crozier), and uncolied in the last three weeks. (J. D. Caponetti and T. A. Steeves. 1963. Can. J. Bot. **41** : 545.)

quantitative differences. Cell division is greatly diminished in the cultured fronds whereas cell enlargement and maturation are essentially normal. This has several consequences for leaf development in vitro. Most of the reduction in duration of development just mentioned occurs in the crozier formation and loosening phase where extensive cell division is involved, and the uncoiling and expansion phase is of almost normal duration. Moreover, in the cultured fronds cell enlargement is not delayed until the final expansion phase of the leaf as in the normal frond but actually participates in leaf growth during crozier formation. What seems to be involved here is a precocious maturation of the leaf when it is isolated from the parent plant. The basic morphogenetic control must remain intact; but the evidence is clear that some aspects of the normal regulation are lacking, and that these have a great deal to do with the size and complexity of the leaf.

Some information is available concerning the nature of the influences that regulate these later stages of development in fern leaves. Some of these clearly are contained within the leaf itself, some give hints of regulation of a rather specific nature by the whole plant, and some suggest rather nonspecific influences of the whole plant system. It has been pointed out that the expansion and maturation phase of leaf development in *Osmunda* does not seem to be adversely influenced by isolation of the leaf. In other studies (Steeves and Briggs, 1960) it has been shown that cell elongation in the rachis, a major aspect of the expansion phase, is promoted by auxin produced in the pinnae during laminar development and transported in a polar fashion out of the pinnae and basipetally along the axis. Removal of the pinnae causes elongation to cease; but it can be restored by application of IAA. Because the uncoiling of the crozier is accomplished only when the wave of elongation extends into the crozier, this process too is stopped by pinna removal and restored by IAA application. Because the mechanism by which this phase is regulated is demonstrated within the leaf, it is not surprising that it is unimpaired by isolation. However, why it begins precociously in excised fronds is an unanswered question.

Because excised leaves complete their development in culture on a medium of simple composition, it has been customary to carry out experiments primarily on such a nutrient supply. Although leaves are photosynthetic organs, they are dependent upon the parent plant for their supply of nutrients, even carbohydrates, during the early stages of development. Long ago Goebel (1908) suggested that carbohydrate supply has an important influence upon leaf form in ferns, and that the small, morphologically simple leaves of the juvenile stage reflect the restricted nutritional status of the plant. In support of these ideas, experiments with excised leaves have shown a marked effect of carbohydrate supply in the medium upon leaf size and form. Caponetti succeeded in doubling the length of the third season fronds of *Osmunda* by raising the concentration of sucrose in the medium from two to six percent, although it was necessary to transfer the leaves to two percent sucrose after the crozier had formed in order to permit uncoiling and expansion to occur (Steeves, 1963). Histological analysis of the larger leaves produced under these conditions showed that the increased size was the result of an enhancement of cell division and that cell length was unaffected. Thus the major defect of isolated leaves was partially overcome by an enhancement of nutrition. Considering that the absorption and translocation of nutrients in an isolated leaf must be greatly restricted in comparison to a normal leaf with intact vascular connections to a parent plant, it seems likely that the reduced cell division activity could be totally explained in this way.

Interestingly enough, the influence of nutrition was even more profound in the case of leaves excised in their second season of development. Sussex and Clutter (1960), by varying the sucrose concentration of the medium, were able to produce mature leaves showing a wide range, not only of size, but of morphological complexity as well (Fig. 9.3). On very low concentrations, small two- or three-lobed leaves were formed that were essentially comparable to those of a juvenile plant. It is difficult to

(a)   (b)

(c)   (d)

Fig. 9.3 Development of excised leaf primordia of *Osmunda cinnamomea* in culture media containing different concentrations of sucrose. Leaves were excised from the plant when they were at the end of the second growing season and 1.6 mm high. (a) Small bilobed leaf that developed on a medium containing 0.006% sucrose. (b) Three-lobed leaf that developed on a medium containing 0.025% sucrose. (c, d) Larger pinnate leaves that developed on media containing 0.5 and 2.0% sucrose respectively. (I. M. Sussex and M. E. Clutter. 1960. Phytomorphology **10**: 87.)

interpret these results because of the problem of evaluating the importance of nutrient reserves in leaves of various ages at the time of excision. Moreover it would be necessary to have histological information of the type presented here for third-season leaves in order to understand fully the developmental responses of younger leaves to nutrient variations. Tentatively, however, the results point to a lessening dependence of leaves upon the parent plant with increasing age and stage of development.

Although it is difficult to make rigid distinctions between nutrients and growth factors, especially when mechanisms of action are not understood, it is important to consider the question of specific hormonal regulation of leaf development in relation to the phenomenon of autonomy. It has been pointed out already that leaves of *Osmunda* and a number of other ferns have been shown to produce auxin in their pinnae, which promotes and is essential for the elongation of the rachis in the expansion phase. This mechanism, however, is leaf-contained and functions effectively in isolated leaves. The evidence for hormonal influences from outside the leaf suggests that they play at most a minor role in later leaf development. In cultured excised leaves addition of various hormones or other growth regulating substances to the medium has been shown to be without effect in a few cases, and more commonly has been inhibitory. An exception to this general statement is the marked promotion of leaf apical growth by kinetin in the presence of adequate sugar (Sussex, 1964). This result seems to indicate that specific growth substances derived from the parent plant may play a role in the regulation of later stages of fern leaf development; but it cannot diminish the overriding importance of carbohydrate nutrition in determining leaf size and shape.

Heteroblastic leaf development

The examples presented thus far have been relatively easy to analyze because they have involved the direct effect of substances upon the leaf growing in isolation. However, it is now necessary to question to what extent the variety of leaf form in an individual fern plant can be explained in terms of the direct action of nutrients within the leaf. It is characteristic of ferns to produce a succession of leaves in the early stages of development of the sporophyte that show a progressive increase in size and complexity. This *heteroblastic* development has interested botanists for many decades as a clear example of the formation of leaves of various forms in a genetically uniform system. Goebel believed that the explanation for the leaf sequence lay in the changing nutritional status of the plant, particularly the carbohydrate level, and Wetmore (1953) has provided experimental confirmation for this. In *Adiantum,*

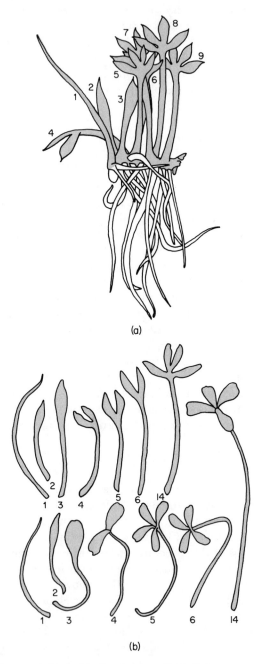

Fig. 9.4   Leaf shape variation in *Marsilea drummondii.* (a) A plant that was grown in sterile culture in a liquid medium containing 3% glucose. The leaves are numbered successively.

sporelings grown on a medium containing sucrose at a concentration of 0.5 percent or higher rarely if ever formed juvenile, two-lobed leaves; and the heteroblastic transition was quickly disposed of. More strikingly, shoot apices of adult plants produced two-lobed leaves if grown on sugar concentrations of 0.1 percent or lower. Wetmore interpreted his observations as indicating a direct carbohydrate influence upon the leaf and felt that the morphologically simple leaf of the young sporeling is the result of an energy deficit that restricts cell division. The results obtained with excised leaves, discussed earlier, tend to support this view.

Working with the aquatic fern *Marsilea* growing in sterile culture, Allsopp (1963) has obtained a similar effect of sugar upon the heteroblastic leaf sequence (Fig. 9.4). On the other hand, Allsopp has found a number of other substances that also influence the sequence and believes that the controls must be somewhat more complex, or at least, less direct. A variety of nutrients and hormones were found to influence the heteroblastic development of leaves in *Marsilea*. The substitution of ammonium ions or of urea for nitrate or the inclusion of gibberellic acid in the medium accelerated the rate of heteroblastic change, as did the addition of casein hydrolysate or certain amino acids. Indoleacetonitrile, which promoted petiole and internodal elongation, reduced the rate of change, and kinetin induced pronounced abnormalities in the leaves which obscured the sequence. In attempting to explain these results, Allsopp has favored an indirect influence of the various substances upon leaf development operating through the whole plant. Recognizing that there is considerable enlargement of the shoot apex during sporeling development, he has suggested that it is apical size that is influenced by the nutritional and hormonal status of the plant, perhaps through direct effects upon protein systhesis, and that this in turn influences developmental process in the leaf primordium in some unspecified manner. In support of this interpretation, Allsopp has shown that several inhibitors of metabolism have a retarding effect upon heteroblastic development or may even cause a reversion to the production of more juvenile leaf forms.

At present the question is unresolved as to whether leaf form is regulated by direct action of nutrients upon or within the leaf as suggested by

---

The first-formed leaves (1–3) are all simple in outline but differ in shape; leaf 4 is bilobed; and leaves 5–9 are four-lobed. (b) Leaves from plants that had been grown in media containing different glucose concentrations. The leaves from each plant are numbered sequentially. The leaves in the upper row are from a plant in 3% glucose, and the transition from simple to two-lobed shape was made at leaf 4, and from two- to four-lobed at leaf 14. The lower row of leaves are from a plant in 5% glucose. The progression through the various shapes is accelerated. Leaf 4 is the first two-lobed leaf, and leaf 5 the first four-lobed leaf. Observe the difference in size and orientation of the pinnae in leaf 14 of the two treatments. (a) ×4, (b) ×3. (A. Allsopp. 1963. J. Linn. Soc., (Bot.) **58**: 417.)

the excised leaf experiments or is influenced indirectly through the parent plant. One particular example will serve to illustrate this problem clearly. Sporelings of the fern *Todea*, if grown in culture on very high concentrations of sucrose, not only form adult-type leaves but also, in many cases, initiate sporangia. These produce spores that are viable and may ultimately be germinated. Interestingly enough, if excised leaves are exposed to the same sugar concentrations in culture, they also initiate sporangia, but the development of sporogenous tissue is arrested at the spore mother cell stage just prior to meiosis and no spores are formed (Sussex and Steeves, 1958). In *Osmunda*, leaves taken from adult plants can also be induced to form sporangia by the same means, and again they are arrested at the spore mother cell stage. If, however, fertile pinnae are removed from fronds developed on the intact plant to the spore mother cell stage, then meiosis will occur and spores are formed (Clutter and Sussex, 1965). Earlier excision leads to a failure of meiosis to occur. It is thus suggested that factors emanating from other parts of the plant can exert an influence upon the occurrence of meiosis.

The leaves of angiosperms

The production of leaves of varied forms by the same plant is commonly observed among flowering plants on which may be found cotyledons, vegetative leaves of various forms, bracts, prophylls, and a variety of specialized organs such as spines, cataphylls, and insectivorous leaves. This variety of forms greatly complicates the problem of interpretation of controlling mechanisms and in addition there is very little precise information obtained from the study of excised leaves. It is therefore not possible to determine to what extent the factors that influence leaf development act directly upon the leaf and to what extent they operate indirectly through the parent plant.

Later leaf development

Before considering what is known of the factors that regulate leaf form in the angiosperms, it is desirable to review a few pertinent facts about the histology of lamina development. At the time of cessation of apical growth in the leaf primordium the peg-like outgrowth is ordinarily dorsiventral but it does not have a lamina (Fig. 9.5). This is initiated by the activity of marginal meristems which appear at, or shortly before, the end of apical growth along the two lateral edges of the primordium. In a simple leaf a strip of marginal meristem, consisting of marginal and submarginal initials, becomes active and proceeds to form the layers of tissue that constitute the leaf lamina. The marginal phase, like the

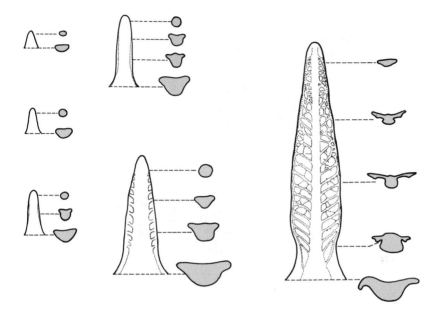

Fig. 9.5 Early stages in the development of a vegetative leaf of *Nicotiana tabacum* (tobacco). At each stage the leaf is shown in face view on the left and in transverse section at several levels. The primordium first develops as a small peg-like outgrowth that is widest at the base. Development of the blade begins later, and because blade development is more intensive above the base the outline of the leaf becomes ovate in face view. (G. S. Avery. 1933. Amer. J. Bot. **20**: 565.)

apical phase, is limited in duration and ceases while the leaf is still very immature and is contained within the apical bud. Often marginal activity ceases first at the apical end of the primordium and the cessation proceeds in a basipetal direction. In the development of compound leaves the process is more complex because lamina formation is initiated in local regions along the peg-like axis, each of which ordinarily goes through an apical and marginal phase of development.

The major size increase of the leaf occurs during a final phase of expansion and maturation during which both cell division and cell enlargement contribute to a great increase in laminar area. The marginal meristems of the leaf and their recent derivatives establish a relatively precise pattern of layers in the unexpanded lamina (Fig. 9.6). As the lamina expands, the general restriction of cell divisions to a plane at right angles to the surface of the leaf, except where vascular tissues are being initiated, maintains the layered pattern. This meristematic activity continues for a surprisingly long time, in some cases until the leaf has attained one-half to three-fourths its final size, before being replaced by

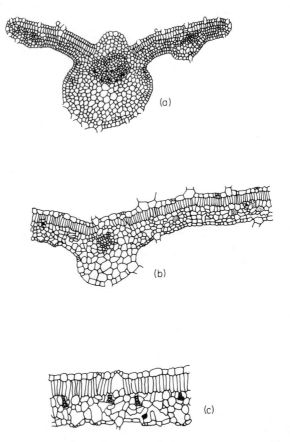

Fig. 9.6 Development of the blade in vegetative leaves of *Nicotiana tabacum* (tobacco). (a) The blade is unexpanded and the precise pattern of layering is visible, especially on the left side. The palisade mesophyll cells form a vertically elongated layer under the upper epidermis. (b) Part of the blade of a leaf that is 150 mm long. Cells are expanding and intercellular spaces are present. The palisade cells are also expanding but they elongate at right angles to the plane of the blade. (c) Part of the blade of a leaf that is 210 mm long and has reached about two thirds of its final size. Palisade cells are tightly packed together, but in the lower spongy mesophyll large intercellular spaces have formed where cells are pulled apart by expansion of the blade. (G. S. Avery. 1933. Amer. J. Bot. **20**: 565.)

net cell enlargement. The layered pattern of laminar growth assumes real importance when it is recognized that the intensity and duration of both cell division and cell enlargement differ in different layers. These relationships, at successive stages of development, have been well documented in *Xanthium pensylvanicum* by Maksymowych (1963). The morphological characteristics of the layers in the mature leaf often reflect these differences, as for example the contrast in organization between palisade and spongy mesophyll. The regulatory mechanisms that maintain control over the development of parallel layers in an expanding lamina must be exceedingly precise, and thus far there is little to indicate their nature. Where a petiole is formed, it ordinarily appears late in the development of the leaf and arises by intercalary growth in the axis of the leaf below the lamina. There is a widespread tendency for leaves to reach maturity first at the apex and subsequently in a basipetal progression, a feature of growth that contrasts with the general acropetal direction of maturation in the stem and is a reflection of the determinate nature of leaf development.

Heterophylly

The occurrence of more than one type of leaf on the same plant, that is, *heterophylly*, fundamentally is genetically determined because the pattern is characteristic of a species. However, the various types of leaves must be induced within the range made possible by the genetic constitution because this presumably remains constant throughout the plant. Although they are not absolutely distinct, two types of patterns of heterophylly may be recognized. In some cases there is a gradual transition of leaf types in the development of the plant, as for example in seedlings in which juvenile forms progressively give way to more adult types. In other cases one stable leaf form gives way rather abruptly to another distinctive leaf type with few or no intermediate stages represented, for example in the formation of bud scales or cataphylls around the terminal bud of a vegetative shoot (Fig. 9.7). In many reported cases the differences in leaf form begin to make their appearance at very early stages in leaf

Fig. 9.7 Leaf types in the terminal bud of *Rhodo-dendron*. External view of the bud of a dormant plant. The outermost cataphylls resemble reduced vegetative leaves and do not sheath the bud. Inner cataphylls sheath the bud and enclose the immature vegetative leaves.

development. Foster (1932) has been able to recognize cataphylls at very early primordial stages in a number of woody species on the basis of distribution of growth in the vertical and tangential planes of the primordia; Fisher (1960) in *Ranunculus hirtus* has been able to detect differences in lobing associated with temperature differences at the P_2 or P_3 stage.

Because of the widespread distribution of heterophylly in angiosperms and its varied expressions, it seems unlikely that any single controlling mechanism can be sought as the explanation. A variety of factors has been shown to influence the expression of heterophylly in a number of cases (Allsopp, 1965). Among factors that have been implicated in different instances are the following: nutrition, light intensity, photoperiod, temperature, water supply, correlative factors within the plant, and various hormones including auxins and gibberellins. Unfortunately among the many studies of factors affecting later leaf growth there are very few that shed any light upon the fundamental question of how the development of the leaf is actually regulated. An exception to this general statement is the careful study of Bostrack and Millington (1962) on leaf form in *Ranunculus flabellaris*, a plant that forms lobed or highly dissected leaves depending upon the environment. Submerged plants produce dissected leaves whereas plants growing terrestrially give rise to expanded leaves (Fig. 9.8). Lobing of the leaf begins at the P_2 stage and almost immediately the higher degree of lobing of the submerged form can be detected. There is no difference in the size or organization of the shoot apex in submerged and terrestrial plants. In addition to submergence,

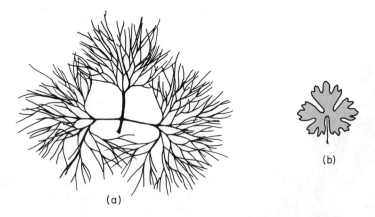

(a)

(b)

Fig. 9.8 Variation of leaf shape in *Ranunculus flabellaris*. (a) Dissected leaves produced on submerged plants. (b) Lobed leaves with expanded laminas produced by plants growing terrestrially. (J. M. Bostrack and W. F. Millington. 1962. Bull. Torrey Bot. Club **89**: 1.)

low temperatures and long photoperiod enhance the development of the dissected leaf type, although in terrestrial conditions long photoperiod favors the expanded leaf form. In order to determine whether these results might be interpreted as direct influences upon the leaf, the apical bud alone was exposed to submerged or subaerial conditions with the remainder of the plant maintained in the alternative conditions. Treatment of the bud alone was as effective as exposure of the whole plant; and the authors have concluded that the influences probably act directly upon the leaf primordia rather than indirectly through the whole plant.

General comment

Although the diversity of observations and the lack of understanding of controlling mechanisms makes any general conclusions about the regulation of later leaf development impossible, a summarizing statement may be useful. The leaf primordium, once determined, manifests a high degree of autonomy or self-regulation in its morphogenesis. It has the capacity to become a leaf rather than some other organ or a disorganized mass if supplied with basic nutrients. It seems also to be able to respond directly to some external stimuli that influence its development. But the leaf is not, in normal circumstances, fully autonomous even though much of its dependence upon the whole plant appears to be rather nonspecific as in the case of nutrients. Highly specific effects produced by certain nutrients normally obtained from the plant are seemingly determined by the leaf itself, and excised leaves respond as well as attached ones. On the other hand, evidence is increasing that there may be more specific stimuli perhaps of a hormonal nature that have their origin elsewhere in the plant, although the best studied hormonal system seems to be leaf-contained. In future work on leaf morphogenesis it might be suggested that the distinction between intrinsic and extrinsic controls is an important one to make, especially as the use of excised leaves in sterile culture greatly facilitates the making of this distinction.

REFERENCES

Allsopp, A. 1963. Morphogenesis in *Marsilea*. J. Linnean Soc. London (Bot.), **58**:417–427.

——. Heteroblastic development in cormophytes. *In* Encyclopedia of Plant Physiology, W. Ruhland, [ed.] Vol. 15(1): 1172–1221. Springer-Verlag. Berlin.

Bostrack, J. M., and W. F. Millington. 1962. On the determination of leaf form in an aquatic heterophyllous species of *Ranunculus*. Bull. Torrey Bot. Club **89**:1–20.

Caponetti, J. D., and T. A. Steeves. 1963. Morphogenetic studies on excised leaves of *Osmunda cinnamomea* L. Morphological studies of leaf development in sterile nutrient culture. Can. J. Botany **41**:545–556.

———. 1970. Morphogenetic studies on excised leaves of *Osmunda cinnamomea* L. Histological studies of leaf development in sterile nutrient culture. Can. J. Botany **48**:1005–1016.

Clutter, M. E., and I. M. Sussex. 1965. Meiosis and sporogenesis in excised fern leaves grown in sterile culture. Bot. Gaz. **126**:72–78.

Fisher, F. J. F. 1960. A discussion of leaf morphogenesis in *Ranunculus hirtus*. New Zealand J. Sci. **3**:685–693.

Foster, A. S. 1932. Investigations on the morphology and comparative history of development of foliar organs. III. Cataphyll and foliage-leaf ontogeny in the black hickory (*Carya buckleyi* var. *arkansana*). Am. J. Botany **19**:75–99.

Goebel, K. 1908. Einleitung in die experimentelle Morphologie der Pflanzen. Teubner, Leipzig and Berlin.

Maksymowych, R. 1963. Cell division and cell elongation in leaf development of *Xanthium pensylvanicum*. Am. J. Botany **50**:891–901.

Steeves, T. A. 1963. Morphogenetic studies of fern leaves. J. Linnean Soc. London (Bot.) **58**:401–415.

Steeves, T. A., and W. R. Briggs. 1960. Morphogenetic studies on *Osmunda cinnamomea* L. The auxin relationships of expanding fronds. J. Exptl. Botany **11**:45–67.

Sussex, I. M. 1964. The permanence of meristems: Developmental organizers or reactors to exogenous stimuli? Brookhaven Symp. Biol. **16**:1–12.

Sussex, I. M., and M. E. Clutter. 1960. A study of the effect of externally supplied sucrose on the morphology of excised fern leaves *in vitro*. Phytomorphology **10**:87–99.

Sussex, I. M., and T. A. Steeves. 1958. Experiments on the control of fertility of fern leaves in sterile culture. Bot. Gaz. **119**:203–208.

Wetmore, R. H. 1953. Carbohydrate supply and leaf development in sporeling ferns. Science **118**:578.

TEN

Shoot Expansion

In preceding chapters emphasis has been placed upon the role of the terminal meristem as the initiating region of the shoot. Clearly it does function as the ultimate source of the cells of the shoot. It is equally true, however, that the meristem proper is by no means the direct source of all or even most of the cells; the multiplication of the immediate derivatives of the meristem in the subapical region amplifies and augments the cellular contribution of the meristem. Nowhere is this more evident than in the phenomena associated with shoot expansion where extensive cell enlargement, coupled with numerous divisions, elaborates the minute structures initiated by the meristem into the recognizable features of the mature shoot. Much of the growth that is of interest to plant physiologists is, in fact, accomplished outside the meristem during the expansion phase of development.

Patterns of shoot expansion

The stem in the vascular plants is a segmented structure, consisting of a sequence of leaf-bearing nodes and internodes. Whatever its ultimate form may be, at the shoot apex the internodes are foreshortened so that the immature leaves are closely crowded around the base of the apical meristem (Fig. 10.1a). In other words, in the shoot apex

139

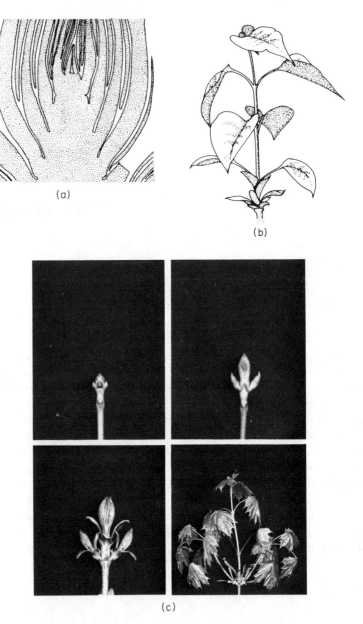

(a)

(b)

(c)

Fig. 10.1   (a) Longitudinal section of the terminal bud of *Syringa vulgaris* (lilac). The internodes are unexpanded. (b) The shoot that results from expansion of a bud such as that in Fig. *a*. (c) Stages in the expansion of buds of *Acer* (maple). (a) × 20, (b) × 0.4. ((a, b) adapted from R. Garrison. 1949. Amer. J. Bot. **36**:205; (c) courtesy of C. Wilson.)

the leaf primordia and the tissues of the stem are initiated but the internodes are not expanded. The expansion of the internodes, which may be extensive or insignificant in different cases, represents a second and relatively distinct phase of development. The nature of this second phase appears most dramatically in the shoots of perennial species in which the two phases are separated by a period of dormancy. In such a plant, if the apical bud is dissected during the period of dormancy, the foreshortened condition of the internodes and the crowding of young leaves around the meristem are readily observed. With the breaking of dormancy, the internodes begin to elongate, the leaves are separated along the extending axis as their subjacent nodes are pushed apart, and the shoot attains its mature form (Fig. 10.1b, c). This process does not involve the most recently formed leaf primordia, which remain as outer leaves in the newly forming terminal bud. In plants whose growth is not interrupted by periods of dormancy, the distinction between the initiation and expansion phases is not so obvious, because internodes at the base of the apical bud expand successively while new leaf primordia and internodes are continually added at the tip.

A shoot that develops in the manner just described is referred to as a *long shoot* (Fig. 10.2a). The use of this term in itself implies the existence of an alternative mode of development and, in fact, *short shoots* are widespread among the vascular plants (Fig. 10.2b). In a short shoot, when the leaves are expanded and mature, the internodes are unelongated so that leaves remain crowded much as in the apical bud. The short shoot habit may be characteristic of a plant throughout its life, as in cycads and many ferns. On the other hand, it may be restricted to certain stages in the life of the plant, giving way to the long shoot habit at a later time, as in the bolting of rosette plants in the reproductive phase. This transition is notable in those plants where elongation is limited to a few, or in some cases apparently to one, internodes followed by a return to the short shoot habit in the reproductive structures. In *Musa* (banana) an infloresence stalk many feet in length consists of only a few internodes, and in *Gerbera* expansion is limited to a single internode. Finally, many plants produce both long shoots and short shoots contemporaneously, a condition seen clearly in *Ginkgo* (Fig. 10.2) and some conifers. The main axes of the plant are long shoots bearing lateral short shoots, and interconversion of the two types can occur under certain conditions. The difference between the two shoot types clearly has its origin in the second or expansion phase of development. If in the maturation of a shoot there is little or no elongation of the short internodes of the apical bud, although there may be considerable increase in the transverse direction, a short shoot will result. In fact, in some perennial shoots the basal internodes of each year's dormant bud do not undergo expansion with the result that over a period of years a

(a)

(b)

Fig. 10.2 Long shoots and short shoots of *Ginkgo biloba*. (a) Long shoots showing the leaves separated by extended internodes. (b) Short shoots with crowded leaves and unextended internodes. (Courtesy of R. H. Wetmore.)

long shoot may be interrupted by periodic zones having short-shoot morphology and indicating growth increments.

The segmental nature of the shoot axis profoundly influences the pattern of growth in the expansion phase, in that each internode appears to develop as a distinctive unit. Thus while the overall growth of the shoot may be observed or plotted, it is also possible to carry out similar analyses of the component internodes. Wetmore and Garrison (1966) have given an account of this kind for the process of shoot elongation in the annual plant *Helianthus annuus* (sunflower). At the time that it is entering the expansion phase, a young internode of *Helianthus* is approximately a millimeter in length. Initially elongation occurs throughout the internode, but as it lengthens, growth is progressively restricted to the upper regions (Fig. 10.3).

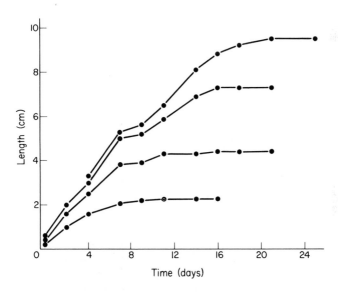

Fig. 10.3 Growth of the first internode of a seedling of *Helianthus annuus* (sunflower). The internode was marked on day 1 into three equal parts when it was 6 mm long. A fourth mark was added at the the top of the internode on day 4 as it elongated. The curves represent the growth of various parts of the internode. (R. H. Wetmore and R. Garrison. 1966. Trends in Plant Morphogenesis. E. G. Cutter [ed.] Longmans, London.)

This was demonstrated by placing evenly spaced India ink marks on one side of the expanding internode when it was long enough to permit this manipulation, and observing the relative spacing of the marks at successive time intervals. Each internode began its expansion slowly, accelerated to a maximum rate of elongation, and then underwent a progressive decline in growth rate. If the length of the internode is graphed against time, the resulting curve corresponds to the classical growth curve. At any one time, most of the growth in the *Helianthus* shoot was concentrated in a single internode, and each successive internode began its most active elongation only when the preceding one was nearing the completion of growth.

Internally it is evident that both cell division and cell enlargement are involved in the elongation of *Helianthus* internodes (Table 10.1).

TABLE 10.1

THE NUMBER OF CELLS AND THE AVERAGE LENGTH OF CELLS IN THE FIRST INTERNODE DURING INTERNODAL GROWTH IN HELIANTHUS. (FROM WETMORE AND GARRISON, 1966.)

Length of internode in mm	Number of cells	Av. length of cells in μ
1.5	77	19.5
2.2	111	19.8
5.1	155	32.9
18.3	267	68.5
78.3	383	204.4
99.5	402	247.5

Wetmore and Garrison found that as an immature internode increased to 65 times its original length there was approximately a five-fold increase in cell number in longitudinal rows in the pith and a 13-fold increase in average cell length. Increase in cell number can be detected in an internode throughout the entire period of growth, but cell division is progressively restricted to the upper regions of the internode. It is evident that no sharp separation between cell division and cell expansion phases is possible in the *Helianthus* internode. In this study attention was devoted only to growth of the ground tissues. Growth in the vascular regions is undoubtedly of a very different type because, during the elongation of the internode, the procambial elements achieve a considerable length. This *symplastic growth*, in which different tissues keep pace in an elongating organ by different mechanisms, is one of the most interesting phenomena of plant development.

Wetmore and Garrison also investigated internodal elongation in *Syringa vulgaris* (lilac), a perennial plant with shoots that, unlike those of *Helianthus*, expand in a flush of growth following a dormant period. Here, although the growth pattern in individual internodes is essentially as in *Helianthus*, there is a marked overlapping of internodal activity, and several contiguous internodes develop contemporaneously although they are not all in the same developmental stage. Another important difference found between the two species was in the relative contributions of cell division and cell enlargement in the ground tissues to the elongation process. In *Syringa* cell numbers in the longitudinal rows increased 65-fold whereas cell length increased only by a factor of three. The enormous difference between these proportions and those in *Helianthus* emphasizes the necessity of investigating the cellular basis of growth in particular cases in conjunction with attempts to elucidate the physiological controls.

Considering the great diversity of vascular plants, it is perhaps not surprising to find tha⁺ there are still other patterns of internodal elongation. In plants such as *Equisetum, Ephedra,* and many monocotyledons, including grasses, the growth sequence in elongating internodes is the reverse of that just discussed. Here the uppermost region of the internode matures first and growth is progressively restricted to the base of the internode. This pattern often results in a persistent growth zone located just above the node that bears the next older leaf, and this zone may function as a persistent isolated meristem bounded both above and below by mature tissues. Such a region often is referred to as an *intercalary meristem.*

Experimental investigation
of shoot expansion

Thus far the phenomenon of internodal elongation has been dealt with descriptively, and the variety of growth patterns by which it is accomplished has been illustrated. In order to understand more fully the significance of this process to the plant and to explore the means by which it is regulated, it is desirable to investigate plants in which the extent of elongation can be modified experimentally. An excellent example of this approach was an investigation of light effects upon the growth of *Pisum sativum* (pea) carried out by Thomson and Miller (1962, 1963). Growth was analyzed in plants exposed to white light or to red light or held in darkness. Significant differences in the heights of plants grown in different light conditions were recorded, and indicated, as expected, that light has a marked inhibitory effect on growth. Analysis of the plants revealed that the reduction in growth could be referred to a shortening of the internodes, that is to a reduction in internodal elongation. The organization of the shoot apex including the apical meristem, the youngest leaf primordia, and their internodes was identical in all treatments. Although the rate of leaf initiation was slower in the dark than in the light treatments, the effect of light was most pronounced on the expansion phase of growth. It exerted an accelerating influence upon both cell division and cell enlargement so that the immature internodes began to expand sooner in the light than in the dark. However, the overall effect of light was to reduce internodal elongation by abbreviating the duration of this phase of development. This resulted from acceleration of the final maturation processes that terminated both cell division and cell enlargement. This study thus testifies to the significance of the expansion phase in determining the final form of the plant body and the extent to which this phase may respond independently to external influences.

The role of the expansion phase of shoot growth may also be studied in certain mutant forms in which the dwarf habit has a genetic basis. In *Lycopersicon esculentum* (tomato) and *Zinnia elegans* Bindloss (1942) found that in a dwarf variety and a tall variety of each there was little difference in the meristem and that the difference in stature had its inception in the subapical region. Dwarfing was caused by reduced mitotic activity correlated with an acceleration of cell elongation. It should be noted, however, that other instances are known in which the activity of the meristem itself is involved in the mechanism of dwarfism (Pelton, 1964).

An obvious next step in the consideration of the expansion phase of shoot development is an analysis of the mechanisms by which it is controlled, that is, of the factors that regulate cell division and cell elongation in the subapical region and their source in the plant. There is a large body of physiological literature dealing with this problem. A complete review of it is beyond the scope of this chapter; it is examined by Galston and Davies in another book in this series.* Here a few instances will be discussed in which the physiological mechanisms have been particularly well correlated with cellular phenomena. In the studies of *Helianthus* described earlier the dependence of stem expansion upon the expanding leaves was explored experimentally. It was found that removal of the pair of leaves at a particular node had a marked depressing effect upon elongation of the subjacent internode so long as the leaves were excised before they were half grown (Fig. 10.4). The effect was sharply limited to a single internode and little influence was noted on development of the internode below or above. Thus the leaf and its subjacent internode seem almost to constitute a growth unit in development of the shoot, a matter of interest in relation to the step pattern of elongation of the axis previously discussed. It is important, however, to note that removal of the pair of leaves did not completely suppress internodal elongation in the subjacent internode, suggesting that the process is not completely dependent upon the associated leaves for the stimuli required for development. Such a response to leaf removal would suggest the operation of a hormonal mechanism, but no direct evidence for this is available in

---

* Galston, A. W., and P. J. Davies. 1970. Control mechanisms in plant development. Prentice-Hall, Englewood Cliffs, N. J.

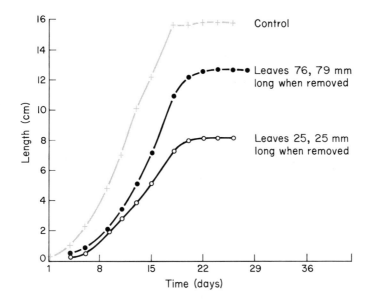

Fig. 10.4   The effect of removing the second pair of leaves on the growth of the second internode of *Helianthus annuus*. The graph shows the growth rate and the final length of the internode of intact plants (+—+), in plants where the second leaf pair was removed on day 10 (●—●) and in plants where the second leaf pair was removed on day 1 (○—○). (R. H. Wetmore and R. Garrison. 1966. Trends in Plant Morphogenesis. E. G. Cutter [ed.], Longmans, London.)

Helianthus. However, in *Coleus*, where the production of auxin by developing leaves has been well documented, Jacobs and Bullwinkel (1953) have shown a similar reduction in internodal elongation following leaf removal, but in this case they demonstrated an almost complete recovery following the application of the hormone indoleacetic acid in concentrations equivalent to that normally produced by attached leaves. Thus the evidence argues for production of auxin by developing leaves having a promoting influence upon elongation in the subjacent internode of the stem.

On the other hand, there are well-documented cases in which the auxin associated with internodal elongation is largely or entirely produced in the stem itself. In *Cercidiphyllum* Titman and Wetmore (1955) found that leaves contributed a negligible amount to the total auxin yield from elongating shoots, and Gunckel and Thimann (1949) had reached a similar conclusion for expanding long shoots of *Ginkgo*. In *Ginkgo*, however, there were indications that the leaves do contribute some factor that is required for auxin production in the stem. This may be an auxin precursor, because removal of the leaves two days prior to testing caused a marked reduction in auxin yield from the stem. Both of these plants produce long and short shoots, and the difference in activity between the two shoot types could be correlated with differences in auxin production in the stem, the pattern being similar in the two species (Fig. 10.5). Both long and short shoots yield diffusible auxin from their

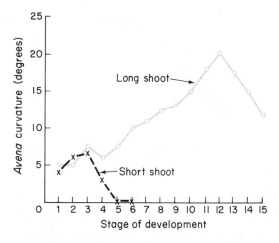

Fig. 10.5 Auxin production by long and short shoots of *Cercidiphyllum japonicum* at successive stages of development in the spring. Auxin yield is expressed in degrees of curvature in the standard *Avena* coleoptile bioassay. (P. W. Titman and R. H. Wetmore. 1955. Amer. J. Bot. **42**: 364.)

terminal buds as they turn green and begin to expand after a period of dormancy, but as the bud opens auxin production falls off to a low level. In the short shoots there is no further yield of auxin even though the attached leaves are expanding. As the long shoots begin to elongate, however, there is a second rise in auxin yield to a level considerably above the earlier level and this occurs in the actively expanding internodes themselves. These several studies emphasize that the dependence of internodal elongation in the stem upon the attached leaves is a variable one, and they also suggest that, though auxin is implicated in the process, it probably is not the sole controlling agent.

This conclusion is supported by recent studies on rosette plants that are converted to long shoots with the onset of flowering. In *Gerbera jamesonii*, for example, Sachs (1968) has found that elongation of the scape or inflorescence stalk is dependent upon the presence of the terminal floral head. If this is excised, scape growth stops but may be partially restored by terminal applications of either indoleacetic acid or gibberellic acid. When these were applied together growth was restored to nearly the values obtained in intact control plants, but histological examination of the ground parenchyma cells revealed that hormone-induced growth had resulted almost entirely from cell elongation and very few divisions had occurred. If, however, only the flowers were removed from the inflorescence, leaving the receptacle and surrounding bracts, combined indoleacetic acid and gibberellic acid treatment again restored scape growth to control levels but cell division was maintained at a normal level and a scape of normal morphology as regards cell number and size was produced.

The role of gibberellins has been investigated in more detail in other rosette plants such as *Samolus* and *Hyoscyamus* (Sachs et al., 1959) (Fig. 10.6). These substances have a marked promoting effect on sub-apical meristematic activity—that is, upon the cell division component

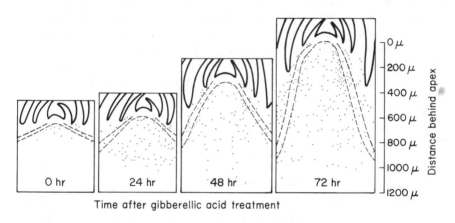

Fig. 10.6 The effect of giberellic acid on subapical meristematic activity in the biennial rosette plant *Hyoscyamus niger*. Mitotic figures, each represented in the figure by a dot, were counted in the median sections of the stem tip at each stage. (R. M. Sachs et al. 1959. Amer. J. Bot. **46** : 376.)

of stem expansion—but they do not influence mitotic activity in the terminal meristem of the shoot. That this action of gibberellins is not restricted to rosette plants is demonstrated by some ingenious experiments on the caulescent plant *Chrysanthemum morifolium* in which treatment with the substance "AMO-1618," which has antigibberellin activity, effectively converted these plants into rosettes (Sachs et al., 1960). The induced rosette habit was then able to be reversed by application of gibberellic acid just as in natural rosette plants.

General comment

Clearly the regulation of stem elongation is not a simple phenomenon and more than one substance must be involved in the process. Much of the difficulty in understanding the control of internodal elongation lies in the fact that both cell division and cell enlargement are involved and undoubtedly are responsive to different stimuli. Auxin produced in the mitotically active tissues of the elongating stem would be expected to promote cell elongation in the same region or in subjacent zones as it moves basipetally through the plant, and in certain cases it is supplemented by supplies from other sources, principally the developing leaves. The role of gibberellins in promoting subapical mitotic activity thus leads to an overlap with the auxin phenomena and to the auxin-gibberellin interaction already cited. Furthermore there is reason to expect that other substances, possibly of the type believed to promote cell division actively in plant tissue cultures, may participate in the expansion phase at least in some cases. These observations pertaining to the regulation of stem elongation again emphasize the distinctiveness of developmental processes occurring in the terminal meristem and in the subapical expanding region.

REFERENCES

Bindloss, E. A. 1942. A developmental analysis of cell length as related to stem length. Am J. Botany **29**:179–188.

Gunckel, J. E., and K. V. Thimann. 1949. Studies of development in long shoots and short shoots of *Ginkgo biloba* L. III. Auxin production in shoot growth. Am. J. Botany **36**:145–151.

Jacobs, W. P., and B. Bullwinkel. 1953. Compensatory growth in *Coleus* shoots. Am. J. Botany **40**:385–392.

Pelton, J. S. 1964. Genetic and morphogenetic studies of angiosperm single-gene dwarfs. Botan. Rev. **30**:479–512.

Sachs, R. M. 1968. Control of intercalary growth in the scape of *Gerbera* by auxin and gibberellic acid. Am. J. Botany **55**:62–68.

Sachs, R. M., C. F. Bretz, and A. Lang. 1959. Shoot histogenesis: the early effects of gibberellin upon stem elongation in two rosette plants. Am. J. Botany **46**:376–384.

Sachs, R. M., A. Lang, C. F. Bretz, and J. Roach. 1960. Shoot histogenesis: subapical meristematic activity in a caulescent plant and the action of gibberellic acid and AMO-1618. Am. J. Botany **47**:260–266.

Thomson, B. F., and P. M. Miller. 1962. The role of light in histogenesis and differentiation in the shoot of *Pisum sativum*. I. The apical region. Am. J. Botany **49**:303–310.

———— 1963. The role of light in histogenesis and differentiation in the shoot of *Pisum sativum*. III. The internode. Am. J. Botany **50**:219–227.

Titman, P. W., and R. H., Wetmore. 1955. The growth of long and short shoots in *Cercidiphyllum*. Am. J. Botany **42**:364–372.

Wetmore, R. H., and R. Garrison. 1966. The morphological ontogeny of the leafy shoot. *In* E. G. Cutter [ed.] Trends in plant morphogenesis. p. 187–199. Longmans, London.

ELEVEN

Modified

Shoot Development

and Flowering

Previous chapters considered how the basic plan of the vascular plant shoot is initiated and elaborated. It will be recalled that shoot development occurs in two relatively distinct phases. An initial phase involves terminal meristem activity in which the tissues and organs are laid down. There follows a phase of expansion growth in the subapical part of the shoot during which the previously formed structures enlarge and mature. Chapter Ten examined how variations in the extent of the expansion phase could produce shoots of widely differing morphology. However, there are other developmental modifications of the basic body plan of the shoot in which the phase of terminal meristem growth is principally involved.

It might be expected that if terminal meristem activity is modified there might be cases of extreme modification in the kind of organs produced and in the extent and pattern of their subsequent growth and development, and indeed this is so, as any student of plant taxonomy or morphology knows. These modifications have been the subject of extensive researches in which the question has been the degree of homology between the modified organs and the more usual organs of the shoot. However, in this chapter attention will be confined to some examples that have proved to be especially amenable to developmental analysis and about which relatively recent information is available.

Determinate meristems—thorns

One of the most striking modifications commonly encountered is the thorn, supposedly protective in function, which is a shoot—terminal or axillary, branched or unbranched—whose apex has ceased to grow and has become more or less hardened and sharply pointed. In terms of developmental modification, the central feature of thorns is that they are determinate organs. In some cases the shoot thorn is little modified from an ordinary vegetative branch. For example, in *Hymenanthera alpina*, a shrub of dry mountainsides, the thorn is scarcely more than an arrested shoot with a blunt tip. Frequent reversion to vegetative growth occurs, but acropetal extension of cork cambial activity ultimately seals off the tip and prevents further development (Arnold, 1959). In *Gleditsia*, on the other hand, Blaser (1956) has shown an interesting transition in the organization of the shoot meristem as it changes from the potentially indeterminate vegetative condition to the determinate state characteristic of thorns. The branched thorns arise ordinarily as axillary buds and initially possess a dome-shaped apex like that of any vegetative shoot. As the transition occurs, there appears to be an acceleration of mitotic activity throughout the meristem and the apex becomes narrower and more elongated. The leaves produced are scale leaves, and two or more of them subtend lateral shoots that also become thorns. As the transformation of the apex begins, leaf production ceases. The acceleration of apical mitotic activity is soon followed by total cessation of divisions, and elongation and differentiation of cells convert the former apex into a narrow pointed thorn. Much of this differentiation produces sclerenchyma causing the thorn to be extremely hard and resistant. An interesting feature of the maturation of the thorn is that tissues remain meristematic in the subapical region to a late stage of development and lead to considerable elongation of the thorn after the tip has become mature.

In *Ulex europaeus*, studied by Bieniek and Millington (1967), almost all lateral shoots are transformed into thorns. The axillary meristem produces only a few scale leaves before undergoing extremely rapid elongation much as in *Gleditsia*. The terminal cells of the attenuating tip elongate and differentiate relatively early and maturation then proceeds basipetally. Sclerification is extensive, especially in the terminal portion of the shoot.

Determinate meristems—
flowers and inflorescences

The relatively simple examples of the transformation of a shoot meristem from the typical vegetative state to a specialized and determinate condition perhaps may serve as a basis for an examination of the more complex, but in many ways comparable, transformation involved in floral development. Reproduction in the vascular plants involves the production of spores that subsequently develop as the sexual, haploid phase of the life cycle, the gametophyte. The production of spores is accomplished in diverse manners in various groups of vascular plants and involves varying degrees of modification of the shoot. Such modifications are perhaps best known in the angiosperms, where they constitute the flowering process and often occur in response to specific environmental conditions such as the photoperiod. Very much less is known about the developmental basis of comparable phenomena in the gymnosperms and the vascular cryptogams although on phylogenetic grounds one might expect to find the underlying physiological mechanisms less complicated by accessory phenomena.

Reproductive morphology

The varying extent to which the shoot system is modified in diverse groups of vascular plants is revealed by consideration of some examples. In the ferns the leaves alone are involved and in many cases a spore-bearing leaf is not significantly different from a vegetative frond. In the horsetails and most clubmosses—members of the *Sphenopsida* and *Lycopsida,* respectively—terminal cones are produced that bring vegetative growth to a close in the shoots on which they occur. In the gymnosperms cone production likewise involves the conversion of an entire shoot to a determinate reproductive structure. In the flowering plants a variety of relationships exists between the individual flower and the shoot system. An individual flower may arise through modification of a single vegetative shoot apex as in tulip. More often a vegetative shoot is transformed into a flowering shoot or inflorescence that contains several or many individual flowers. In such cases the flowers often arise directly from the inflorescence meristem without a preceding vegetative phase. Reproductive development of the angiosperms, like that of other vascular plants, is generally characterized by determinate growth of the meristems involved, but there are well-known cases, both normal and abnormal, in which there is a return to indeterminate vegetative growth.

Development of the floral apex

Where a single flower is produced it is possible to follow the transformation of a vegetative shoot apex into a reproductive axis in a relatively uncomplicated situation. This transition has been described by Engard (1944) in *Rubus rosaefolius*, (Fig. 11.1), the vegetative apex of which is a low-domed mound with a tunica usually of three layers, but with some variation in the number of layers. With the onset of flowering the apical mound begins to enlarge by a dome-like expansion that

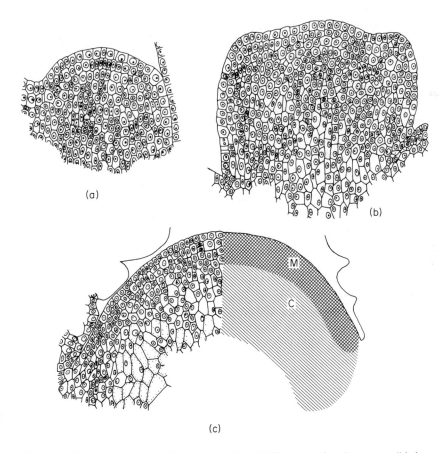

Fig. 11.1 Floral development in *Rubus rosaefolius*. (a) The vegetative shoot apex. (b) An early stage in the development of a flower. The two outgrowths on the sides of the meristem are the sepal primordia. (c) Later stage of flower development showing the structure of the meristem in which can be detected a superficial mantle and an underlying core. Key: C, core; M, mantle. ×325. (C. J. Engard, 1944. Univ. Hawaii Res. Public. **21** : 1.)

appears to result from two processes. There is an acceleration of cell division in the upper cells of the corpus in the center of the apex, with the divisions oriented predominantly in the periclinal plane. Coupled with this is an extension of vacuolation and cell enlargement acropetally from the maturing pith region into the apex. Subsequently, cells of the tunica layers also begin to divide actively, and ultimately the apical dome, now much expanded, consists of a shallow mantle of small meristematic cells stretched over a central parenchymatous core. Cell division is uniformly active throughout the mantle and only the outermost tunica layer has retained its identity through the restriction of divisions to an anticlinal plane.

As these changes in growth pattern occur in the apical mound, there are correlated alterations in lateral appendage production. The first five primordia, the initials of the sepals, are formed in rapid succession with essentially no internodal spacing between the successive primordia. This tendency to rapid initiation of appendages with little spacing is characteristic of flower development, and is seen very clearly in the development of the flower of *Aquilegia formosa* as described by Tepfer (1953) (Fig. 11.2). As in *Rubus* the first changes associated with the onset of reproductive development are an increase in height and width of the apex brought on by increased cell division and an expansion of the parenchymatous core. Five sepals are initiated separately in rapid sequence, followed by five petals that are established simultaneously and in positions that alternate with those of the sepals. The 40 or 45 stamens are then initiated in simultaneous whorls of five, followed by two whorls of petaloid staminodia. Finally five carpels are formed at the summit of the apex. The acropetal succession of appendage initiation proceeds relatively more rapidly than the growth of the apical mound with the result that the apex is progressively restricted in size, ultimately remaining as a small, flat, and differentiated plate of tissue in the midst of the carpels. Thus at the time that the flower is fully formed, there are no cells remaining that retain the characteristics of a meristem, and the shoot is determinate. The initiation and early development of all of the lateral appendages in *Aquilegia* showed sufficient similarity to the development of leaves to support the conclusion that they are fundamentally foliar in nature. From the phylogenetic point of view this is in accord with the interpretation of floral appendages held by many workers. On the other hand, in more highly specialized flowers the departure of some appendages from leaf morphology and developmental pattern is so great that one may well question the value, or even the validity, of seeking homologies with leaves.

The development of *Michelia fuscata*, which was described by Tucker (1960), is of interest because of the prolonged activity of the floral

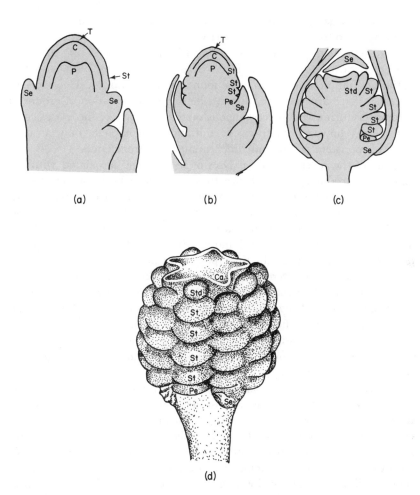

Fig. 11.2 Development of the flower of *Aquilegia formosa*. (a–c) Longitudinal sections of flowers at different developmental stages showing the emergence of successive whorls of lateral appendages. (d) External appearance of a flower at the stage comparable to that shown in *c* with the sepals removed so that the younger primordia can be seen. The petals are very small. Above them are several alternating whorls of stamens and staminodia. The uppermost whorl of lateral organs, the carpels, is just emerging. Key: C, corpus; Ca, carpel; P, parenchymatous core; Pe, petal; Se, sepal; St, stamen; Std, staminodium; T, tunica. a–c ×75. (a–c. S. S. Tepfer, 1953. Univ. Calif. Public. Botany **25**: 513; d. S. S. Tepfer et al., 1963. Amer. J. Bot. **50**: 1035.)

meristem, the great ultimate length of the floral axis, and the large number of appendages produced in this primitive flower. In the extended growth of the floral meristem, although the overall pattern of change in the apex is similar to that described in the two previous cases, there are periodic fluctuations in size comparable to those occurring in a vegetative apex, so that a greater similarity between vegetative and reproductive development is revealed. Plastochronic changes in size are characteristic of vegetative apices where they are associated with leaf initiation but are not ordinarily found in floral apices. However, in *Michelia*, there are periodic intervals between the initiation of groups of primordia during which conspicuous increases in size of the apex occur. Ultimately, however, after carpel initiation is complete, only a small plate of meristem is left that differentiates as parenchyma.

Some of the most careful analyses of the changes associated with the onset of reproductive growth are those that have been carried out on plants in which there is precise photoperiodic control of flowering. In such cases it is possible to follow the time course of floral initiation and development with considerable accuracy. An excellent example of this approach is the analysis by Wetmore, Gifford, and Green (1959) of the development of the reproductive apex in *Xanthium pensylvanicum*, a plant that has been used extensively in photoperiodic studies. In contrast to the cases previously described the vegetative apex in *Xanthium* is converted to an inflorescence, a compact head containing many individual flowers, rather than to a single flower (Fig. 11.3). This species is extremely sensitive to the photoperiodic stimulus and some varieites will respond by flowering after exposure to only a single short-day photocycle, although the response is more rapid and complete if several cycles are given. The vegetative apex has the typical angiosperm tunica-corpus organization and also reveals the presence of a distinct central zone of enlarged, lightly staining, and apparently mitotically sluggish cells. The first visible change in the apex is an acceleration of division at the base of the central zone just above the rib meristem. These newly formed cells, together with those of the rib meristem, soon begin to enlarge, and the combination of cell division and cell enlargement leads to a swelling of the apex, which becomes distinctly mounded in external view. Meanwhile cells in the overlying part of the central zone have undergone anticlinal divisions, and with the continuation of division in the cells of the peripheral zone this results in the development of a small-celled meristematic mantle covering an expanded parenchymatous core. On this expanded dome lateral primordia are initiated in an acropetal sequence, but these are not floral organs as in the previous cases. Rather they are the primordia of bracts in the axils of which small individual flowers are produced.

This examination of the formation of reproductive apices in several species by transformation of vegetative meristems has revealed the rather surprising fact that as far as the nature of the transformation is concerned it makes little difference whether the resulting structure is to be a single flower or an inflorescence. Wetmore, Gifford, and Green have emphasized

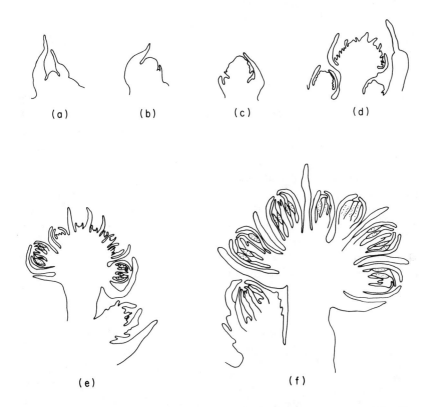

Fig. 11.3 Changes in the shoot apex during inflorescence development in *Xanthium pensylvanicum*. (a) The vegetative apex showing the meristem and youngest leaf primordia. (b–f) Stages in the development of the meristem as an inflorescence with the differentiation of numerous flowers. (F. L. Naylor, 1941. Bot. Gaz. **103** : 146.)

this point in their comparative study in which the essential features of the transformation were the same in the compact head of *Xanthium*, the branching inflorescence of *Chenopodium album*, and the single flower of *Papaver somniferum*. It is also worthy of note that the first two of these are short-day plants whereas the last is a long-day species. Although the changes associated with the onset of flowering are ordinarily regarded as representing reproductive development, it seems reasonable to ask whether, in fact, the fundamental change is not associated with the onset of determinate growth. In this connection the similarity between the changes in developing flowers and inflorescences and those that occur in thorns, determinate shoots having nothing to do with reproduction, is of considerable interest. In all these cases there appears to be an initial acceleration of mitotic activity that heralds the end of meristematic development in the apex. Observations such as this do not give support to the méristème d'attente hypothesis of apical organization, discussed in Chapter Five, in which the central region is considered to be a meristem specifically destined to produce the reproductive parts of the flower (Buvat, 1952).

Although there have been many histological descriptions of the change from vegetative to reproductive development in shoots, it is only recently that attention has been turned to the biochemical changes occurring within the cells of a transforming meristem. Much of the cytochemical work to date has been summarized by Nougarède (1967) in an extensive review of meristem structure and function. Though the studies thus far completed are not extensive enough to warrant generalizations, a picture has begun to emerge of rather fundamental changes in the synthetic activities of meristem cells following the initiation of reproductive development. In view of the acceleration of mitotic activity, it is not surprising that labeling experiments reveal increased synthesis of DNA, which is particularly notable in the central zone (Fig. 11.4). It is interesting to note, however, that in *Sinapis alba*, which can be induced to flower by

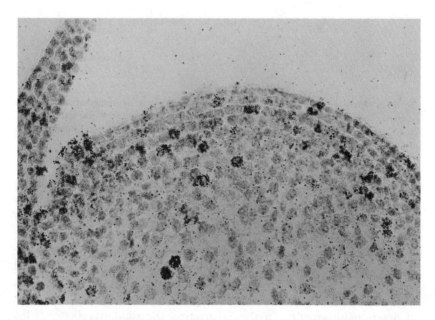

Fig. 11.4 An early stage of inflorescence development in *Helianthus annuus* (sunflower). The shoot tip has been labelled with tritiated thymidine and autoradiographed. Compare with Fig. 5.2, the corresponding treatment of a vegetative apex, and note the uniform distribution of labelled nuclei in the central and the peripheral regions of the inflorescence meristem. ×260. (T. A. Steeves et al., 1969. Canadian J. Bot. **47** : 1367.)

a single long-day photocycle, the rise in mitotic activity precedes the increase in DNA synthesis by some hours (Bernier, 1969). This indicates the presence in the meristem of cells which are held in the stage of the cell cycle immediately following DNA synthesis and are thus ready to divide without further DNA synthesis. Associated with increased DNA synthesis is a corresponding increase in synthesis of RNA, revealed by autoradiography, which results in an increase in the content of RNA, particularly in the cytoplasm, shown by specific staining techniques. Furthermore, in contrast to its localized distribution in the vegetative apex, RNA is rather uniformly distributed throughout the meristematic mantle of the reproductive apex. Shortly after the increase in RNA there is a rise in the total protein content of the cells of the meristem, which may be a further biochemical step following from the increase in RNA. Gifford and Tepper (1962) have shown in *Chenopodium album* that at a relatively late stage in inflorescence initiation there is a decrease in histone content of the nuclei in the meristem as indicated by a reduced staining reaction. Subsequently Gifford (1964) found in *Xanthium* that three days after the inductive photocycle the cytoplasm of the meristem cells showed a staining reaction characteristic of histones.

These studies point to progressive change of the terminal meristem with the onset of reproductive development. The distinct zonation of the vegetative apex gives way to the uniform appearance of the reproductive meristem, not because the floral apex originates from only a specific region, such as a méristème d'attente reserved for reproductive development, but because there is an increased rate of mitotic activity in the cells situated at the tip of the meristem.

Experimental analysis of floral meristems

The reproductive apex with its predictable sequence of appendages contrasts sharply with the repetitive pattern of growth exhibited by the vegetative apex. Several workers have applied the same sort of experimental methods used in the study of the vegetative apex to elucidation of the reproductive apex. Not surprisingly the results have often differed. No attempt will be made here to explore the vast literature dealing with the production, translocation, and action of the flowering stimulus. Rather, attention will be focused on the experiments that deal with the reproductive apex as a developing system and seek to understand its organization and the interactions among its parts.

It is evident that in the initiation of reproductive development, a vegetative shoot apex responds dramatically to a stimulus. Although the response in this case involves an entire shoot rather than a single organ, it may be reasonable to ask whether the induction of a flowering apex is not comparable to the determination of a leaf primordium. In both cases a group of meristematic cells capable of producing a vegetative shoot are altered in their development to the extent that they produce an entirely different structure. In both cases the altered meristem proceeds through a series of well-defined stages to complete maturation. Experiments discussed in Chapters Eight and Nine have shown that early in its development a leaf primordium is determined and subsequent to this is relatively autonomous in completing its developmental destiny. If plants of *Chenopodium album* are subjected to two inductive short-day cycles they will flower even if they are returned to long days for the remainder of their development. This would suggest that some sort of determination occurs, but it is necessary to distinguish clearly between the induction of a plant to flower and the actual determination of one or more of its meristems. The necessity for such a distinction is particularly clear in plants that continue to grow vegetatively even while they are flowering. Although the plant as a whole is induced to flower, only some of the apices are determined as floral meristems; the remainder continue development as vegetative meristems.

In *Chenopodium*, Wetmore, Gifford, and Green have specifically tested the determination of the individual apex by excising it from the plant six days after the two inductive cycles were given and growing it in isolation in sterile culture. Under these conditions excised apices continued to develop and produced small, but seemingly normal, inflorescences. Thus there seems to be a stage of development in the reproductive apex, as in the leaf, beyond which no further external stimulus or specific control is required. Unfortunately, information of this sort is not available for other species so that it is impossible to know whether this is a general phenomenon. However, in *Aquilegia* (Tepfer et al., 1963), it was found that floral meristems excised at the time sepals were being formed were able to initiate all of the floral organs in sterile culture although the later stages involving flower maturation did not occur normally.

Some observations on intact plants also indicate the existence of a determination process. In *Chrysanthemum*, Schwabe (1959) has described experiments in which inflorescence development in this short-day plant is arrested by transfer to long days, by reducing light intensity, or by applying auxin. The significant point regarding determination was that if a particular apex had passed a critical stage, indicated in this case by the development of the carpels in the marginal flowers of the inflorescence, it continued its development regardless of the fate of other apices of the same plant. The apices that had not reached the critical stage at the time of removal from inductive conditions did not revert to the vegetative condition but expanded into enlarged receptacle-like, determinate shoot tips and were arrested in their development. If the arrested apices remained viable, subsequent return to inducing conditions brought about renewed activity and the completion of inflorescence development. Thus it is possible, as Popham and Chan (1952) have suggested, that determination is a two-step process, with determination of the receptacle preceding the stage in which flower initiation is induced. This system would be an extremely favorable one for the application of excision and culture techniques such as have been employed for *Chenopodium*, for in this way the developmental capacity of the meristem at successive stages could be assessed accurately.

In contrast to these observations which suggest that determination manifests a high degree of permanence in a meristem, mention should be made of some other cases in which there appears to be natural or experimentally induced reversion to the vegetative state in meristems that had entered upon reproductive development. Although meristem reversion has been rather frequently reported for plants bearing apices in the earliest transition stages, a remarkable case of reversion of a well-developed floral meristem was reported in *Impatiens balsamina* by Krishnamoorthy and Nanda (1968). When plants were subjected to four or more inductive short days to induce flowering and then returned to long days, all of the meristems could revert to the vegetative condition even after the formation of several or all the whorls of floral organs was completed. If there is a developmental stage in this plant that could be called determination it appears that it is very labile. In view of this result considerable interest is attached to the surgical experiments of Wardlaw (1963) on inflorescence meristems of *Petasites hybridus*. Working with a series of apices in the transition from the vegetative to the flowering condition, Wardlaw variously punctured them at the center, bisected them, or isolated the meristem from adjacent subapical tissues by vertical incisions. When these operations were carried out on relatively advanced stages, inflorescence development continued from portions of the original apical surface, indicating a determination in the meristem. When the experiments were performed on early transition stages, however, there was a more or less complete return to the vegetative condition. Thus determination of the reproductive meristem becomes progressively more fixed as development proceeds. An interesting aspect of these experiments was that direct application of solutions of gibberellic acid to the regenerating apex significantly favored the retention and further elaboration of reproductive development.

The reproductive apex passes through a series of relatively distinct morphological stages during which lateral organs of different types are produced. Surgical experiments by Cusick (1956) on flowers of *Primula bulleyana* have attempted to explore the physiological basis of this succession (Fig. 11.5). At various stages in their development the floral apices

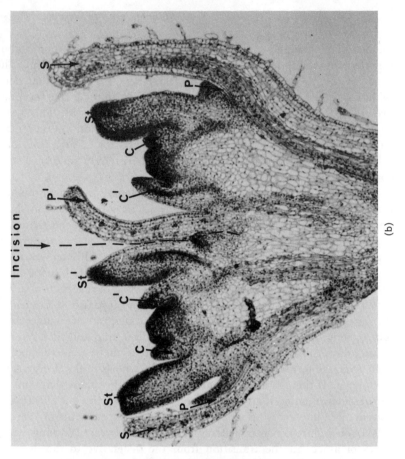

(b)

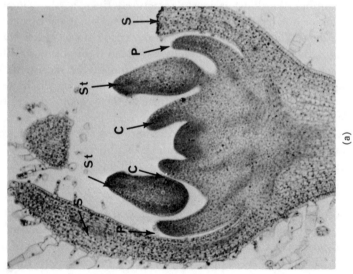

(a)

164

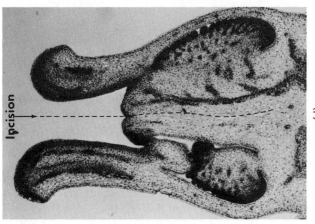

(d)

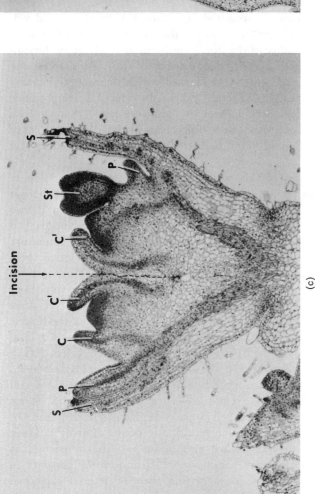

(c)

Fig. 11.5   Surgical bisection of the floral meristem of *Nicotiana tabacum* (tobacco). Results obtained on this species were generally similar to those obtained on *Primula*. (a) Longitudinal section through an intact apex after all the floral organs have been initiated. (b–d) Development of floral meristems that had been vertically bisected, then excised from the plant and grown in sterile culture. (b) Meristem bisected before sepal primordia were initiated. On the outer side of each flower organ formation has proceeded normally. On the inner face of each flower, where organ regeneration has occurred, stamens, petals, and carpels have been initiated in their correct positional relationships, but no sepals have been formed. (c) Meristem bisected after sepals and petals had been initiated. The ability to regenerate stamens and petals on the inner face of each flower has been lost, but carpels are still regenerated. (d) Floral meristem bisected at the early carpel stage of development. All regenerative capacity has been lost, but the two halves of the flower have grafted together along the incision. In *d* only the inner part of the flower is shown in the photograph. Key: C, carpel; P, petal; S, sepal; St, stamen. The preceding are the normal organs that formed on the outer side of the flower. C′, regenerated carpels; P′, regenerated petals; St′, regenerated stamens. (a) ×70, (b) ×35, (c) ×28, (d) ×21. (G. S. Hicks and I. M. Sussex, 1971. Bot. Gaz., in press.)

165

were bisected, and particular attention was paid to the nature of the appendages produced by the regenerating half-apices. If the meristem was split at the early presepal stage—that is, when the meristem was recognizably larger than a vegetative apex but before the sepal primordia or any other primordia had been initiated—each half-apex regenerated a complete flower with all appendages present. If, however, the developing flowers were split at midpoint in the presepal stage, no sepals were formed on the side of the regenerated flowers above the incision, although sepals were formed normally on the intact side of the apex. At the late presepal stage the operation resulted in absence of stamens, petals, and sepals on the operated side of the regenerated flowers. Bisections at stages after initiation of sepals also resulted in failure of normal ovary development. These experiments emphasize the progressive restriction of developmental potentiality of the floral meristem as it advances through the sequence of stages, a restriction associated with the determinate nature of the flower and contrasting markedly with the repetitive development of the indeterminate vegetative shoot. In view of the importance of this concept of floral development, it should be noted that Cusick's findings have been confirmed, with differences in detail, for flowers of *Portulaca grandiflora* (Soetiarto and Ball, 1969) and of tobacco (Hicks and Sussex, 1971).

Experimental results of this sort immediately raise the question of possible interactions among the various parts of the flower, and there is some experimental evidence to suggest that floral appendages do influence the development of other appendages. This problem has been approached in an interesting way in the flowers of *Aquilegia* by isolating immature floral buds and culturing them on a nutrient medium. Under these conditions Tepfer and his associates detected antagonistic interactions among various appendages. The sepals were found to exert an inhibitory influence on the development of all other floral parts and, in fact, must be removed to obtain satisfactory growth of the remainder of the flower. Further, the growth of stamens was poor under most conditions, but improved considerably if the carpels were destroyed or inhibited. In this connection it may be noted that the different floral organs responded distinctively to hormones incorporated in the medium. It was found that gibberellic acid stimulated the growth of all floral organs except the stamens, and that the carpels were especially sensitive to IAA, too low or too high a concentration resulting in failure of growth. These distinctive hormonal responses are also of interest in comparison to Cusick's report of progressive restriction of organ-forming capacity in the developing floral meristem, suggesting that the altered capacities may result from a changing hormonal status in the meristem. The possible significance of the hormonal status of the floral meristem in the orderly sequence of development is also suggested by the often drastic modifications of floral morphology induced by treatment with growth-active substances (Vieth, 1965).

Of particular interest among the examples of organ interaction in floral development are those that have been discovered in studies of unisexual flowers. In *Cucumis sativus* there are three types of flowers produced, male, female, and hermaphroditic, depending upon the genetic constitution of the plant and the growth conditions. They are indistinguishable in their early development, and if excised from the plant at this time and grown in sterile culture it is possible to convert potentially male flowers into female flowers by including IAA in the culture medium (Galun et al., 1963). Gibberellic acid was found to nullify the auxin effect. In no case did flowers in culture become hermaphroditic, suggesting that there is an antagonism between the development of stamens and of carpels, and no treatment was found that would convert potentially female flowers into male flowers.

Heslop-Harrison (1964) has utilized results such as these and others that he obtained by studying sex reversions in other plants to formulate a theory of flower development involving specific interactions between the organs of successive whorls. He supposes that substances released by the first-formed whorl activate genes that lead to the development of the next type of organs, and so on. Though where is no information on the nature of substances that might be involved in such interactions, this approach provides a useful background against which to view experiments in which floral development has been modified.

General comment

In the development of thorns and of reproductive apices, the characteristic zonation of the vegetative apex disappears and is replaced by a uniformly meristematic surface, and this is accomplished particularly by changes in the central zone including an acceleration of mitotic activity. So striking are the changes in the central zone that one group of investigators (Buvat, 1952) regarded this region as a meristem that specifically initiates the reproductive portions of the flower. Careful time-course studies, however, have shown that the whole vegetative apex is involved in the transformation to the reproductive state, and the similarities of the changes in apices being converted to thorns and to reproductive meristems argue against such an interpretation. It has already been pointed out that reorganization of the meristem in flowering seems related more to the onset of determinate growth than to reproductive activity specifically; and this then points to the probable significance of the central zone as a mechanism that serves to maintain continued meristematic activity of the vegetative apex.

It has been seen in Chapter Five that although there are several different interpretations of the functional organization of the vegetative shoot apex, the idea is gaining favor that fundamental importance should be attached to a central group of cells known variously as the central zone, the metrameristem, or the méristème d'attente. In addition to its distinctive cytochemical characteristics this group of cells is considered to be relatively sluggish in mitotic activity in comparison with the more active peripheral regions of the meristem. It may be reasonable to suggest that a region characterized by a high rate of mitosis is unable to maintain this activity indefinitely, so that rapid division is the forerunner of ultimate differentiation. Thus the shoot apex is provided with a reservoir of centrally located cells from which its differentiating regions are continually supplied and which remain meristematic by virtue of the fact that they do not divide frequently. Conditions that lead to increased mitotic activity of these cells result in determination of the meristem, which then may proceed through an elaborate succession of distinctive stages but which ultimately becomes fully differentiated and nonmeristematic.

REFERENCES

Arnold, B. C. 1959. The structure of spines of *Hymenanthera alpina*. Phytomorphology **9**:367–371.

Bernier, G. 1969. *Sinapis alba* L. In: The induction of flowering. Some case histories. L. T. Evans [ed.], pp. 305–327. Macmillan, Melbourne.

Bieniek, M. E., and W. F. Millington, 1967. Differentiation of lateral shoots as thorns in *Ulex europaeus*. Am. J. Botany **54**:61–70.

Blaser, H. W. 1956. Morphology of the determinate thorn-shoots of *Gleditsia*. Am. J. Botany **43**:22–28.

Buvat, R. 1952. Structure, évolution et fonctionnement du méristème apical de quelques dicotylédones. Ann. Sci. Nat. Ser. 11 **13**:199–300.

Cusick, F. 1956. Studies of floral morphogenesis. I. Median bisections of flower primordia in *Primula bulleyana* Forrest. Trans. Roy. Soc. Edinburgh **63**:153–166.

Engard, C. J. 1944. Organogenesis in *Rubus*. Univ. Hawaii Research Pub. **21**:1–234.

Galun, E., Y. Jung, and A. Lang. 1963. Morphogenesis of floral buds of cucumber cultured *in vitro*. Devel. Biology **6**:370–387.

Gifford, E. M., Jr. 1964. Developmental studies of vegetative and floral meristems. Brookhaven Symp. Biol. **16**:126–137.

Gifford, E. M., Jr., and H. B. Tepper. 1962. Histochemical and autoradiographic studies of floral induction in *Chenopodium album*. Am. J. Botany **49:**706–714.

Helsop-Harrison, J. 1964. Sex expression in flowering plants. Brookhaven Symp. Biol. **16:**109–125.

Hicks, G. S., and I. M. Sussex. 1971. Organ regeneration in sterile culture after median bisection of the flower primordia of *Nicotiana tabacum*. Bot. Gaz. **132:**(in press).

Krishnamoorthy, H. N., and K. K. Nanda. 1968. Floral bud reversion in *Impatiens balsamina* under non-inductive photoperiods. Planta. **80:**43–51.

Nougarède, A. 1967. Experimental cytology of the shoot apical cells during vegetative growth and flowering. Internat. Rev. Cytol. **21:**203–351.

Popham, R. A. 1964. Developmental studies of flowering. Brookhaven Symp. Biol. **16:**138–156.

Popham, R. A., and A. P. Chan. 1952. Origin and development of the receptacle of *Chrysanthemum morifolium*. Am. J. Botany **39:**329–339.

Schwabe, W. W. 1959. Some effects of environment and hormone treatment on reproductive morphogenesis in the Chrysanthemum. J. Linnean Soc. London (Bot.) **56:**254–261.

Soetiarto, S. R., and E. Ball. 1969. Ontogenetical and experimental studies of the floral apex of *Portulaca grandiflora*. 2. Bisection of the meristem in successive stages. Can. J. Botany **47:**1067–1076.

Tepfer, S. S. 1953. Floral anatomy and ontogeny in *Aquilegia formosa* var. *truncata* and *Ranunculus repens*. Univ. Cal. Pub. Botany **25:**513–647.

Tepfer, S. S., R. I. Greyson, W. R. Craig, and J. L. Hindman. 1963. *In vitro* culture of floral buds of *Aquilegia*. Am. J. Botany **50:**1035–1045.

Tucker, S. C. 1960. Ontogeny of the floral apex of *Michelia fuscata*. Am. J. Botany **47:**266–277.

Vieth, J. 1965. Étude morphologique et anatomique de morphoses induites par voie chemique sur quelques Dipsacacées. Thèse. Dijon.

Wardlaw, C. W. 1963. Experimental investigations of floral morphogenesis in *Petasites hybridus*. Nature **198:**560–561.

Wetmore, R. H., E. M. Gifford, Jr., and M. C. Green. 1959. Development of vegetative and floral buds. *In* R. B. Withrow [ed.], Photoperiodism and related phenomena in plants and animals. Amer. Assn. Adv. Sci. Publ. **55:**255–273.

TWELVE

The Root

The previous eight chapters have dealt with postembryonic development of the shoot. Consideration must now be given to the subsequent development of the other meristem initiated in the embryo, the root apical meristem. The organ system that develops from this meristem during the ontogeny of the plant is as extensive as the shoot system and in many cases exceeds the aerial system in size. Moreover, root systems show considerable morphological diversity and are by no means stereotyped in form and in development. Unfortunately, however, this is not generally appreciated because root systems are inaccessible to direct and sequential observation of the type that can easily be made on the shoot. Without elaborate excavation, root systems cannot be studied except in special cases, and even when exposed can hardly be observed ontogenetically in anything approaching normal circumstances. It is, therefore, regrettable, but not surprising, that much of our knowledge of root development is based upon laboratory-cultured seedlings of annual crop plants.

The remarkable extent of certain individual root systems has been revealed by excavation and measurement. Ecologists have long recognized that in a plant community there is usually a stratification of root systems comparable to the multiple stories of shoots. Such a layering tends to minimize competition among species for water and nutrients in the soil, and it has been shown to have an important bearing upon survival under adverse conditions. For example, in the North American prairies during the great drought, species with root penetration exceeding five feet were greatly favored over more shallow-rooted species. What is perhaps more significant in the present context is that such studies have also demonstrated morphological complexity in root systems. Root systems often consist of both deeply penetrating roots and shallower horizontally extending roots. The horizontal roots may rather abruptly turn downward and assume a vertical direction of growth (Fig. 12.1). Differentiation into vigorously growing long roots and lateral short roots, which are limited and often evanescent, also appears to be a common phenomenon. Such morphological complexity suggests the existence of correlative mechanisms perhaps comparable to those found in the shoot, but little

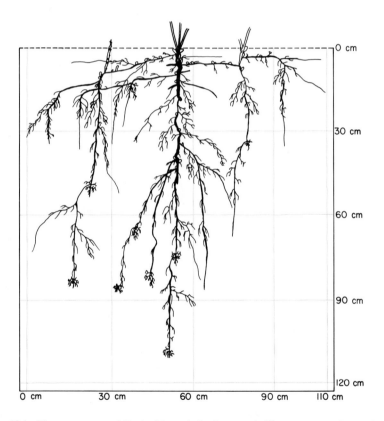

Fig. 12.1 The root system of *Euphorbia esula* (leafy spurge). The root system is morphologically complex. Some roots penetrate vertically downward; others grow horizontally and turn down at differing distances from their point of origin. Horizontally growing roots may give rise to buds which grow upward as shoots. Some lateral roots branch profusely and remain stunted in growth. These are distinguished as short roots. (M.V. S. Raju et al., 1963. Canadian J. Bot. **41** : 579.)

is known of these mechanisms in roots. For example, there is evidence indicating that the long and short root pattern reflects the operation of a kind of apical dominance, but the hormonal or other basis of the phenomenon remains obscure.

The individual root differs from the shoot in certain developmental features, and these must be recognized in any analysis of root growth. The apical meristem of the root is covered distally by a cap of mature tissue and thus is subterminal rather than terminal as in the case of the shoot apical meristem. Consequently the meristem is surrounded by its differentiating derivatives. Moreover the root meristem does not produce any lateral appendages comparable to leaf primordia, and the segmentation both in mature structure and in differentiation of the stem, which is related to the leaves, is lacking in the root. Thus processes of cellular enlargement and maturation can be traced basipetally from the root meristem in an unbroken sequence. In principle, the root ought to be a simpler system in which to study growth processes; but this expectation has been realized only in part.

The root apex

Structure

In the shoot apex the distal surface of the meristem is superficial and is thus clearly delimited, and the lateral boundaries are indicated by the positions of the leaf primordia. Only basipetally, where maturing derivatives abut on the meristem, is there a problem of delimitation. In the root, however, where the meristem is surrounded by maturing derivatives, precise delimitation of the meristem is much more of a problem. This difficulty becomes evident when median longitudinal sections of root apices are examined. In such preparations longitudinal files of mature and maturing cells may be traced acropetally as they converge upon a small region, just below the root cap, whose cell rows also converge upon this region. Because in many cases these files of cells can be traced into a small number of meristematic layers in the root tip the idea developed that the meristem consists of histogenic layers, each of which specifically initiates one or more tissues of the mature root. Such a histogen concept was at one time widely accepted for shoot apices; but whereas subsequent work forced the abandonment of this view for the shoot, it has persisted in descriptive accounts of the root apex.

The application of the histogen concept in the description of root meristems may be illustrated in the case of radish (Fig. 12.2). Here the root meristem consists of three superposed, transversely-oriented tiers of

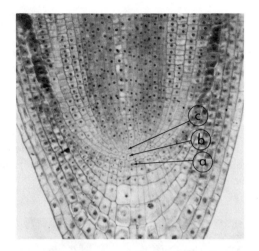

Fig. 12.2   Longitudinal section of the root tip of *Raphanus sativus* (radish). The three tiers of initials in the meristem are labelled *a*, *b*, and *c*. Derivative cells of tier *a* differentiate as root cap and epidermis, those of tier *b* as the cortex, and the derivatives of tier *c* differentiate as the vascular tissue of the root. ×68.

initials. The most distal of these by periclinal and anticlinal divisions gives rise to the root cap and to the epidermis. The middle tier initiates the cortex, and the innermost layer produces the vascular system. Thus the entire meristem consists of cells whose initiating function can be related to specific mature tissues of the root body. This is in contrast to the situation in the shoot apex where tissue differentiation has been described only among the derivatives of the promeristem.

Several other types of root apical organization have been described in which histogens are found, but in which the pattern differs from that just described. For example the root cap may arise from a separate layer, and the epidermis may have a separate origin or arise from the same layer as the cortex. Comprehensive classification schemes for root meristems have been proposed on the basis of number of histogenic layers and their supposed function, but little would be added to this discussion by reviewing these.

The occurrence of superposed meristem layers in the root apex has not always been interpreted as implying the existence of true histogens. For example in *Euphorbia esula* (leafy spurge) relatively distinct tiers have been found in the meristem (Raju et al., 1964), but the number varies from one to four in different roots, and there is some variation in the relationship to mature tissues. These tiers were simply designated as *meristem layers* in *E. esula* and it was concluded that their relationship to certain mature tissues reflected regularity of segmentation patterns in the apex rather than tissue specificity.

The root apices of some angiosperms and many gymnosperms cannot be described in terms of discrete meristem layers but rather they seem to contain a transversely-oriented common initiating region for all tissues of the root. In some cases the transverse meristem has been interpreted as consisting of two plates of initials, one specifically for the vascular system and one for the rest of the root. In other cases, such as *Pisum* (Popham, 1955), cell rows can be traced from the central part of the root cap through the meristem to the vascular tissue; and it is argued that there can be only one set of initials (Fig. 12.3).

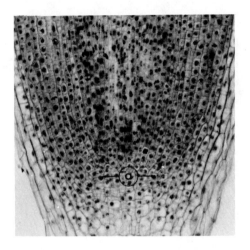

Fig. 12.3   Longitudinal section of the root tip of *Pisum sativum* (pea). Only a single layer of initials (*a*) can be identified in the meristem, and all the tissues of the root are derived from them. ×68.

The concept of common initials for the entire root has been developed along another line as an alternative interpretation of root apical organization. In many ferns and other lower vascular plants there is an enlarged and conspicuous apical cell present in the root, as in the shoot (Fig. 12.4). Through its segmentation along four cutting faces, the apical cell functions as the ultimate initial of all the tissues of the root. Derivatives of the distal face mature as root cap cells while those from the lateral cutting faces initiate the body of the root. Attempts have been made to describe a similar kind of segmentation pattern based upon a single apical initial in the root apices of several angiosperms (von Guttenberg, 1964). It is held that this central cell serves to renew the initials of the actual histogens

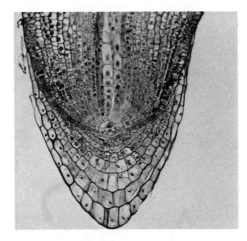

Fig. 12.4 Longitudinal section of the root tip of *Pteridium aquilinum* (bracken). A single enlarged apical cell occupies the center of the meristem. The most recent derivatives of the apical cell are visible as segments around its internal faces. These differentiate as all the tissues of the root. ×68.

where these are present; but in some cases, particularly that of the vascular system, the histogens are believed to be essentially autonomous. In such cases the central cell would not be a complete apical initial. This view of the root apex has found little support from other workers and has had little influence upon studies of root development.

Experimental evidence

A somewhat modified version of the central cell theory postulates, on the basis of experimental evidence, that there may be a small group of apical initials to which all of the tissues of the root may be traced. Brumfield (1943) X-irradiated seedling roots of *Vicia faba* and *Crepis capillaris*, and after a period of growth sectioned the roots and searched for patterns of distribution of X-ray-induced chromosomal aberrations. Where such aberrations were found in dividing cells they occurred in sectors occupying approximately one third of the root and including all root tissues from the center to the surface. Such sectorial chimeras led Brumfield to postulate the existence of three initial cells, each of which ultimately gives rise to all of the tissues of a sector of the root. This conclusion has been criticized on the grounds that the results might reflect the regeneration of a root apex from a small number of cells left viable after the radiation treatment and might have little relationship to normal apical structure. Popham has also suggested that the initiating region in the seedling roots used in this experiment might consist of a very small number of cells so that the results of the experiment might have limited significance.

The questions raised by these divergent points of view concerning the size and organization of the initiating region of the root have been approached in other experimental ways, notably by Clowes, using surgical methods (1953, 1954). He made a series of glancing or wedge-shaped cuts into the root tip at various depths, in *Vicia, Fagus, Zea,* and *Triticum* (Fig. 12.5). It was argued that if there is a single initial cell, or a small group of initials, a cut deep enough to remove this cell or group should prevent further normal development. A slightly shallower incision that leaves the initial cell or group intact should have little effect on development. If, on the other hand, the initiating region consists of a rather large group of cells, both deep and shallow incisions should permit the continuation of root development, probably with abnormalities in the regenerated portion and in the mature tissues produced by it. The difference between deep and shallow incisions would be in the proportion of abnormal to normal meristem after regeneration. In fact the second of these two alternatives was realized in all cases, suggesting that there is no central cell or small group of apical initials having special significance. A further important observation was made in the roots in which distinct histogens could be noted (*Zea* and *Triticum*). In the undamaged portion of the root apex the normal meristem layers could be observed and their relationships to specific mature tissues was as expected. In the regenerated portions of the same apices these relationships had been altered, particularly in the lack of distinctness of a root cap histogen which is a conspicuous feature of root apices of grasses (Fig. 12.5d). It is thus strongly suggested, because all of the root tissues were being produced by the altered meri-

Fig. 12.5 Surgical experiments that examine the size and organization of the root meristem. (a–c) Diagrams that show three types of experiments performed on the root tip. The incisions remove different regions of the terminal meristem. The incision in *a* removes an oblique flank of the meristem; the incision in *b* removes a vertical wedge; and that in *c* removes a horizontal wedge. (d) Median longitudinal section of a root tip of *Zea mays* (corn) fixed eight days after an operation to the left side of the apex of the type shown in figures *a* or *b*. On the right side of the root the normal relationship between cell layers in the meristem and in the differentiating tissues has been maintained. The epidermis and the cortex on this side of the root are shown limited by the dark boundaries that trace their origin into the meristem. The root cap originates from a distinct cell layer in the meristem on this side of the root. On the left side of the root regeneration following the operation has obliterated the normal cell relationships. A distinct root cap initiating cell layer in the meristem is not evident, and files of cells extending from the meristem toward the surface provide evidence of extensive cell proliferation. One such file of cells is shaded in the diagram, and others are visible near it. Key: C, cortex; E, epidermis; RC, root cap. (d) ×250. (a–c. F. A. L. Clowes, 1953. New Phytol. **52**: 48; d. F. A. L. Clowes, 1954. New Phytol. **53**: 108.)

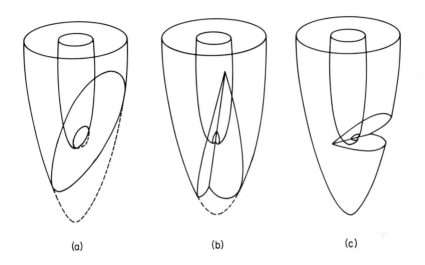

(a)   (b)   (c)

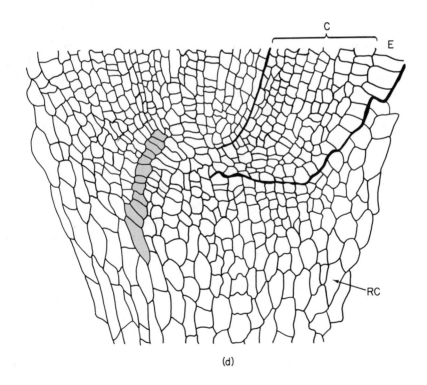

(d)

stem, that the apparent histogens are not tissue specific and therefore are not histogens in the original sense of the word.

The quiescent center

The analysis of root meristem organization has been further complicated but greatly enlivened by the discovery that the very cells in the root apex over which most of the controversy has raged are at least relatively inactive mitotically and, in fact, may not divide at all. In his analysis of the root of *Zea*, Clowes encountered great difficulty in interpreting cellular configurations in the apex and concluded that the observed patterns could exist only if the most central cells of the meristem divide very infrequently or not at all. This rather startling conclusion was strikingly verified when this same investigator (1956a and b) supplied radioactive precursors of nucleic acids to roots of *Zea*, *Vicia*, and *Allium* and used autoradiographic techniques to determine the localization of nucleic acid synthesis in the root meristems. He was able to show conclusively that there is a *quiescent center* in the meristem in which little or no DNA synthesis occurs and presumably little or no mitosis (Fig. 12.6). Other workers have confirmed these results by means of autoradiography as well as by other kinds of observations. Workers of the French school who had made mitotic counts on shoot apices reported a region similar to the méristème d'attente in the apices of roots in which mitotic figures were rare or absent. In onion root tips Jensen and Kavaljian (1958) demonstrated by mitotic counts that there is an essentially nondividing central region in the apex.

In subsequent studies other characteristics of the quiescent center in certain species have been elaborated. The content of RNA and the rate of synthesis of both RNA and protein are lower in the quiescent center cells than in immediately surrounding regions (Clowes, 1956a, 1958b). Such reports are in agreement with the observed fainter staining of cells in this region with ordinary histological stains. Further distinctive features of cells of the quiescent center have been revealed by electron microscopy (Clowes and Juniper, 1964). Differences were noted in several cell organelles, the most conspicuous being the decreased number of Golgi bodies in the quiescent cells and the relatively poor development of mitochondrial cristae as compared with the cells of the surrounding mitotically active regions.

If the cells in the center of the root meristem are largely or completely inactive, one may well ask what is the significance of the often highly precise cellular patterns of the meristem layers that other workers have called histogens. Clowes has provided an answer to this question by a consideration of root development in the early stages of seed germination

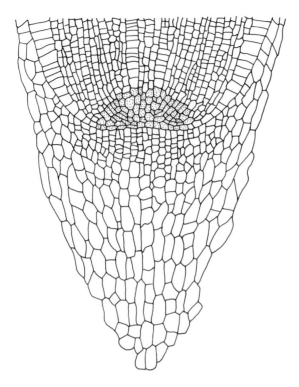

Fig. 12.6   Median longitudinal section of the root tip of *Zea mays* (corn). The quiescent center is shown as stippled cells. The relationship between differentiating cells and tissues and the meristem is very clear in this section. ×160. (F. A. L. Clowes, 1959. Biol. Rev. **34**: 501.)

and in lateral roots. In such cases he has found (1958a) that there often is no quiescent center at the beginning of development and that it appears as growth continues. Thus he is able to argue that cellular patterns in the quiescent center, although they may have no functional significance at a particular moment, reflect the developmental history of the region.

It is apparent that the quiescent center of roots is remarkably like the méristème d'attente claimed to exist in the shoot apex. Interestingly enough, whereas the proposal for the shoot evoked immediate opposition, prolonged debate, and the refutation of some of its original assertions, the quiescent center of roots has been accepted generally and has been demonstrated in many species by a number of workers. It must not be concluded, however, that the existence of nondividing cells in the center of the root meristem poses any fewer problems than would the existence of such a region in the shoot apex. In particular it raises the problem of the identity of the permanently meristematic cells or the promeristem. Clowes has dealt with this problem by locating the initials of the root just outside the quiescent center and completely surrounding it, and he has designated this group of initial cells the promeristem of the root. But Clowes's own analysis of mitotic frequencies in various parts of the root apex (1961, 1962) indicates that cells of the quiescent center do divide, though very slowly. In *Zea* roots, for example, he reports that the cell cycle in the quiescent center has a duration of approximately 170 hours whereas for rootcap initials the duration is 12 hours. In other regions of the root apex the duration of the cell cycle is somewhat greater than in the cap initials, but much less than in the quiescent center.

Unfortunately Clowes has never described the location within the quiescent center of cells that may be seen to divide or to synthesize DNA. Thus it is not clear whether the entire quiescent center is characterized by a slow rate of division or whether divisions occur only in the peripheral region, with the centermost cells remaining inactive. If only the peripheral cells are active mitotically, one might think of the quiescent center as fluctuating slightly in position from time to time, a condition Clowes has suggested. If, on the other hand, even occasional divisions occur throughout the quiescent center, serving to renew the more actively dividing regions around it, then it is difficult to refrain from considering the cells of the center as the initials of the root, and thus as the promeristem. The suggestion of Paolillo and Gifford (1961) that the true initials of the shoot need divide only very infrequently in comparison with surrounding cells in order to fulfill their function applies equally to the root. The latter interpretation, in fact, agrees with the views expressed by von Guttenberg (1964) regarding the organization of the root apex. He recognized that cell divisions occur much more frequently in a peripheral zone surrounding a central group of relatively less active cells, but he regarded the peripheral cells as rapidly dividing derivatives of the true initials. His postulated central cell or cells would of course lie within the relatively inactive group.

There are other observations on root apices that have a bearing on these questions. In the root system of *Euphorbia esula* (Raju et al., 1964) it was discovered that, whereas the indeterminate *long roots* ordinarily showed a distinct quiescent center in terms of DNA synthesis (Fig. 12.7), the determinate laterals, which have been designated *short roots*, lack such a center throughout their development. Clowes has suggested that very narrow roots may never develop a quiescent center in their ontogeny, but such considerations do not appear to apply in *Euphorbia* because

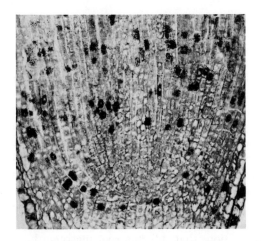

Fig. 12.7   Pattern of thymidine labelling of nuclei in the tip of a long root of *Euphorbia esula* (leafy spurge). Labelled nuclei appear dark because of the accumulation of silver grains over them. Nuclei are labelled near the bottom of the figure where the root cap is initiated, and further back in the root where the other root tissues are initiated. Between these two zones of labelled nuclei is a large zone of unlabelled nuclei which is the quiescent center. ×234. (M. V. S. Raju et al., 1964. Canadian J. Bot. **42**: 1615.)

narrow primary roots of seedlings, which are of the long-root type, consistently showed a quiescent center in their apices.

It would not be difficult to visualize differing degrees of activity in the permanent initials of roots of different types, but it seems illogical to regard the initials as being completely different in long and short roots. Furthermore in *E. esula* there is evidence of a seasonal fluctuation in the size, and perhaps even the presence, of a quiescent center. In these perennial roots, the center is very conspicuous at the height of the growing season; but earlier in the season when growth is being reactivated, the quiescent center is extremely small and in fact may be absent. Clowes and Stewart (1967) have found a similar participation of the quiescent center in the recovery of roots from artificially cold-induced dormancy in seedlings. This suggests that it may be unwise to treat the quiescent center as a morphological entity, as some workers appear to do, and that it is more meaningful to deal with quiescence as a phenomenon that is variably superimposed upon the cellular pattern of the root apex. This view is in complete accord with von Guttenberg's concept of periodic renewal of the peripheral actively dividing regions of the root meristem by the central initiating cells.

Experimental investigation
of root development

As in the shoot apex, it is possible to learn something about functional organization of the root apex using experimental procedures in which normal developmental processes are disturbed. In view of the often stated simplicity of the root, it is surprising to find that relatively little experimental work of this type has been done. It must be pointed out, however, that, in addition to the difficulties of handling roots and root systems, the fact that the root meristem is enclosed makes the use of surgical methods somewhat awkward because the meristem is not visible and cannot be exposed. Nevertheless a few workers have extracted important information about the root by these means.

Longitudinal incisions that split the root meristem have been carried out by Ball (1956) on the mature embryo of *Ginkgo* and by Pellegrini (1957) on *Phaseolus* (bean) seedlings and the results have been the same in both cases. Each side of the split meristem underwent reorganization and produced a complete root. Ball has reported that where the division was unequal, two roots of unequal size were formed. It therefore must be concluded that as in the shoot, a portion of the root meristem has the capacity to form a complete meristem. No experiments have been carried out that indicate the minimum size of a piece of meristem having this capacity.

Clowes's surgical experiments, which were discussed previously for their bearing upon the central cell theory and the specificity of meristem layers, may be reconsidered in the present context. In these experiments a portion of the meristem was excised and the remainder, like the half of a bisected apex, reorganized into a complete meristem. Another type of meristem regeneration has also been demonstrated by Clowes (1959) in response to damaging doses of X-irradiation. In some cases in which radiation causes a cessation of mitosis in the more actively dividing cells of the meristem, the cells of the quiescent center, which seem to be less radiosensitive, began to synthesize DNA actively and to undergo mitosis. These activated cells permitted the continuation of root growth. Although there is no direct evidence, it may be suggested that in the surgical splitting and excision experiments something of the same sort may take place. Both Clowes and von Guttenberg have suggested that the existence of a relatively inactive central group of cells might reflect some kind of competition for nutrients among cells of the meristem. If this is the case, upsetting the normal relationships by surgery or radiation could easily result in increased activity in the central group of cells. It is interesting to speculate as to whether some mechanism involving competition for metabolites might underlie the functional organization of the meristem in which small portions are clearly capable of organizing whole meristems if suitably isolated but do not do so within the intact apex.

It has long been recognized that if a root is decapitated in such a way as to remove all of the terminal meristem and a minimal amount of additional tissue, the stump is often able to regenerate a new meristem so that root growth continues. This response is in contrast to that previously described where the root tip was split longitudinally in that the regenerating cells are not part of the apical meristem but are in fact partially differentiated derivatives. It is also in contrast to the behavior of a shoot tip where the complete removal of the meristem precludes the direct regeneration of a new apex. Torrey (1957) has investigated the course of apex regeneration following decapitation in roots of *Pisum* grown in sterile nutrient culture. The experiments were carried out by removing 0.5 millimeter from the root tip including the cap. The cut passed through a region of the root just proximal to the meristem at which level the vascular tissue was in an early procambial stage of differentiation but no mature elements were present. Cell division in the procambial tissue led to the formation of an outgrowth that soon became organized as a root meristem. The pericycle contributed most actively to the outgrowth and the cortical tissues were not involved at all. If there was considerable damage at the cut surface, ordinary lateral root formation occurred instead of direct apical regeneration. The contrasting behavior of roots and shoots following removal of the meristem may have its explanation in terms of the normal method of branching in the two systems. In the shoot, lateral branches ordinarily arise from the terminal meristem, whereas in the root, lateral members have their origin in mature or partially mature regions from a particular tissue, the *pericycle*, that retains the capacity to produce them. Thus in a decapitated root a new apex may arise in the absence of the terminal meristem, and it is worthy of note that the pericycle plays a leading role in the regenerative process.

The fact that a decapitated root is able to regenerate a new meristem might suggest that the root apical meristem is to a large extent under the control of mature tissues and in this sense is less autonomous than the shoot apex. Great interest is therefore attached to the experiments of Torrey (1954) and Reinhard (1954) in which it has been shown that a root tip essentially of the size removed in the previously reported decapitation experiments is capable, in the case of *Pisum* roots, of forming a root when placed in sterile nutrient culture. As in the case of the shoot, the nutritional requirements for such small explants were rather exacting and half millimeter tips failed to grow on a medium that would support the growth of larger explants. However, the additional requirements for organic nutrients such as vitamins were entirely ones of concentration, and the only new substances required were inorganic salts of the micro-nutrient category.

These experiments point to the essential autonomy of the root meristem; but they do not establish it conclusively, because the smallest explants capable of organized growth include differentiating tissues of root cap and root axis in which patterns of organization are already established. Smaller explants produce only callus but subsequently may give rise to root primordia. Torrey (1955) has provided additional evidence from the same experiments supporting the concept of apical autonomy. He has reported that although the majority of the roots that developed from 0.5 millimeter tips have a vascular pattern like that of the original root, about 20 percent of the tips produced a different pattern that may change during subsequent growth. He concludes, therefore, that the root meristem must control the pattern of tissue differentiation.

Root branching

With the exception of certain lower vascular plants, chiefly the lycopods, branching in the root is a phenomenon that does not directly involve the apical meristem. In the vast majority of vascular plants lateral roots have their origin in the pericycle, one or more layers of cells that constitute the outermost region of the vascular system. Thus lateral roots do not arise from superficially placed buds as do shoot branches; but rather have an endogenous origin that necessitates their subsequent penetration of cortical and dermal tissues. Observation of lateral root primordia is thus as difficult as observation of the terminal meristem of the root. A root primordium is initiated by divisions in a localized group of pericycle cells which give rise to a hemispherical mound of meristematic tissue. (Fig. 12.8) Continued oriented divisions lead to the establishment of a terminal meristem and root cap well before the young root has penetrated to the exterior of the parent root.

Although all the cells of the pericycle presumably are equally capable of initiating a root, it is an interesting observation that lateral roots are spaced with some degree of regularity. The most obvious pattern is one that is related to the internal organization of the parent root. Lateral roots are ordinarily distributed in rows or orthostichies opposite the pro-toxylem poles of the vascular system, with one or two rows opposite each pole. Other patterns, such as roots situated opposite the phloem poles, are known. The mechanism regulating this placement of roots is completely unknown. Apart from study of the stigmarian rootlets of extinct lepidodendrids and the roots of *Isoetes*, little attention has been devoted to aspects of the pattern of lateral root distribution other than the relationship of vertical rows to internal organization and the restriction of laterals to a certain distance from a growing tip. Recently Riopel (1966) has found in the root of *Musa* (banana), where there are numerous orthostichies, that, starting with a particular lateral root, there is a significant probability that the next root to arise will be located in an orthostichy other than that of the reference root. Riopel postulated that this spacing might result from competition for nutrients along the orthostichy. In any event it is evident that the spacing of lateral roots is considerably less precise than the placement of lateral organs, leaves and buds, in the shoot.

It has been observed also that the initiation of lateral roots occurs at a predictable distance from the tip of the growing root at a level at which elongation has ceased, although this distance may vary with the conditions of growth. This might suggest either that a certain level of tissue maturity is required before root initiation can begin or that the main root apex exerts some regulatory influence upon the process of lateral root initiation. The second of these suggestions has attracted more interest and has been the object of several experimental studies. A number of workers have shown that removal of the root tip promotes the formation of lateral roots. Torrey (1952) found that removal of the cotyledons in *Pisum* seedlings grown in sterile culture prevented lateral root formation on the primary root. Thus, it appears that the initiation of laterals is inhibited by the root tip and promoted by more mature regions. Presumably the position at which a new lateral root is initiated is determined by the interaction of stimuli arising in these two locations. Although the identity of neither of these stimuli is fully known, it is apparent that both are complex. It has been found that in isolated roots growing in culture, initiation of laterals may be limited by the availability of auxin, adenine, thiamine, nicotinic acid, or certain micronutrient elements, any of which could therefore act as a promoting substance. At the same time effective inhibitors of lateral root formation have been extracted from *Pisum* roots. The spacing of lateral roots is thus a complex phenomenon in which interaction of many substances appears to be involved.

There is a small amount of evidence suggesting that a similar mechanism may underlie the development of heterorhizic root systems in which laterals may develop as either long or short roots. When the apex of a long root producing only short root laterals ceases growth or is removed, new long root laterals are often initiated that replace the main axis. This suggests a controlling influence of the main root apex upon the growth potential of subjacent lateral roots.

General comment

This treatment of developmental processes in the root has been relatively brief compared to that devoted to the shoot and its component

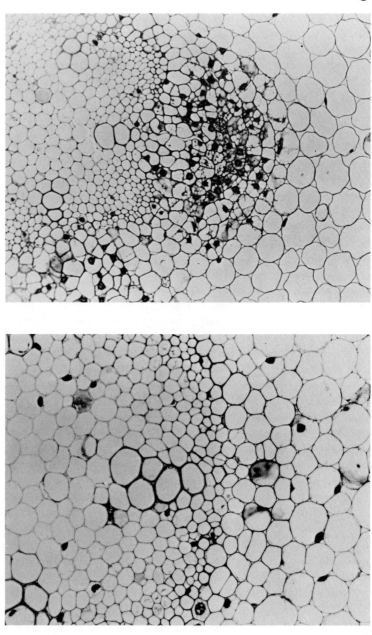

(a)

186

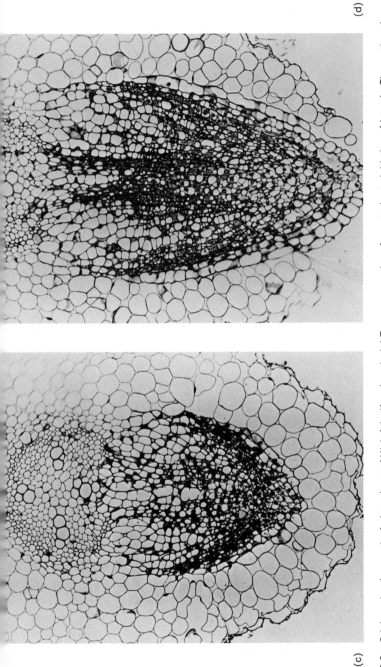

(c)

(d)

Fig. 12.8   Origin and early growth of a lateral root of *Vicia faba* (fava bean).   (a) Transverse section of a root prior to initiation of laterals. The section shows the region where the cortex (the large cells in the lower part of the section) abuts on the vascular tissue above. In the vascular system one of the xylem strands is visible, and lateral roots are initiated just outside this. (b) An early stage in lateral root development. Cells immediately outside the xylem have begun to divide and are starting to organize into the lateral root meristem. Further outside, cortical cells are undergoing an increase in the density of cytoplasm and are also starting to divide. (c) The lateral root is growing through the cortex of the parent root at right angles to the direction of the main root axis.   (d) The lateral root has just reached the surface of the parent root and it has a well organized meristem and root cap. Behind the lateral root meristem vascular tissue, cotex, and epidermis can be seen differentiating. (a) ×432, (b) ×192, (c, d) ×136. (Photographs courtesy H. T. Bonnett.)

parts. This brevity of treatment is possible largely because the root and shoot systems, in spite of their seemingly great differences, have much in common in their basic organization. Both represent continually expanding systems that carry out a three-dimensional exploration of the environment, although the environments are rather different. In both systems the form depends largely upon the activity of the terminal meristems and of the recent derivatives of these meristems, not only for the continued or indefinite growth pattern but for the kind and degree of branching as well.

Although the apices of root and shoot appear rather dissimilar in organization, the differences may be more apparent than real. For example, the root of many species seems to show a close relationship between discrete layers in the meristem and specific tissues of the mature root. The interpretation of these layers as histogens, however, has lost much of its former support; they are now considered to reflect regular patterns of cell division in a meristem that is not disturbed by the periodic initiation of appendages as in the shoot. Perhaps more striking is the seeming comparability of root and shoot apices in the possession of a central zone that is quiescent, or relatively so, as far as cell division is concerned. Although this feature of organization is more widely accepted for the root than for the shoot, its presence in some form in the shoot apex is gaining support. There is no need to suppose that the quiescent center must be identical in shoot and root, but its presence in some form in both would give strong support to the idea that it plays a fundamental role in the functioning of terminal meristems. It may not be too soon to begin to think of such a reservoir of relatively inactive cells as fundamental to the indeterminate or potentially unlimited growth of the terminal meristems. Finally, it may be noted that experiments have shown that both shoot and root apices are to a large extent autonomous or self-determining in their development. These formative centers are differentiated in the early development of the embryo, and they subsequently appear to produce their respective systems with only basic nutritional support from the already mature portions of the plant. Although this last statement is not absolutely proven, experimental evidence increasingly points to its validity.

There are, however, important differences between developmental processes in shoot and root that must be noted. Whereas the shoot meristem is located terminally in the shoot, the root meristem is subterminal as a result of its production of mature derivatives acropetally as well as basipetally. The functional significance of this difference is obvious, but its bearing upon the organization of the root and shoot apex should perhaps be considered. In the shoot apex, the surface is extremely important, and only portions of meristem with the ordinary surface layers present can regenerate whole apices. In the root, on the other hand, this limitation does not apply; decapitated roots can regenerate new apices. The appendages of the shoot, whether leaves or buds, have their origin at the surface and at the margin of the meristem. In the root, however, there is no appendage formation at the surface or near the meristem, and indeed there could not be because any protuberance would be destroyed as the apex is forced through the soil by subapical elongation, or would impede that necessary elongation. Thus lateral roots are initiated at some distance from the apex, by the activation of cells of the pericycle, and at some distance within the tissue of the root. The absence of appendages at the apex results in a much greater regularity of developmental processes within the root apex, and ultimately in the development of a nonsegmented axis in contrast to the nodal-internodal organization of the stem. This has important consequences for the processes of elongation and differentiation.

Thus there are significant differences between the developmental processes in roots and shoots, but reflection upon the even greater similarities leads to the conclusion that these differences are superimposed upon a fundamentally homologous plan of organization. If the original land vascular plants were rootless, as is commonly supposed—that is, if the plant body was not differentiated into root and shoot systems as in more recent vascular plants—then these two systems probably represent evolutionary modifications of an original organizational plan in relation to two different environments. The fundamental similarities of root and shoot apices may well reflect the common origin, and the differences may reflect the evolutionary adaptations to contrasting environments.

REFERENCES

Ball, E. 1956. Growth of the embryo of *Ginkgo biloba* under experimental conditions. II. Effects of a longitudinal split in the tip of the hypocotyl. Am. J. Botany **43**:802–810.

Brumfield, R. T. 1943. Cell-lineage studies in root meristems by means of chromosome rearrangements induced by X-rays. Am. J. Botany **30**:101–110.

Clowes, F. A. L. 1953. The cytogenerative centre in roots with broad columellas. New Phytol. **52**:48–57.

———— 1954. The promeristem and the minimal constructional centre in grass root apices. New Phytol. **53**:108–116.

———— 1956a. Nucleic acids in root apical meristems of *Zea*. New Phytol **55**:29–34.

———— 1956b. Localization of nucleic acid synthesis in root meristems. J. Exptl. Botany **7**:307–312.

—— 1958a. Development of quiescent centres in root meristems. New Phytol. **57**:85–88.

—— 1958b. Protein synthesis in root meristems. J. Exptl. Botany **9**:229–238.

—— 1959. Reorganization of root apices after irradiation. Ann. Botany London [N. S.] **23**:205–210.

—— 1961a. Duration of the mitotic cycle in a meristem. J. Exptl. Botany **12**: 283–293.

—— 1962. Rates of mitosis in a partially synchronous meristem. New Phytol. **61**:111–118.

Clowes, F. A. L., and B. E. Juniper. 1964. The fine structure of the quiescent centre and neighbouring tissues in root meristems. J. Exptl. Botany. **15**:622–630.

Clowes, F. A. L., and H. E. Stewart. 1967. Recovery from dormancy in roots. New Phytol. **66**:115–123.

von Guttenberg, H. 1964. Die Entwicklung der Wurzel. Phytomorphology **14**:265–287.

Jensen, W. A., and L. G. Kavaljian. 1958. An analysis of cell morphology and the periodicity of division in the root tip of *Allium cepa*. Am. J. Botany **45**: 365–372.

Paolillo, D. J., Jr., and E. M. Gifford, Jr. 1961. Plastochronic changes and the concept of apical initials in *Ephedra altissima*. Am. J. Botany **48**:8–16.

Pellegrini, O. 1957. Esperimentic chirurgici sul comportamento del meristema radicale di *Phaseolus vulgaris* L. Delpinoa. **10**:187–199.

Popham, R. A. 1955. Zonation of primary and lateral root apices of *Pisum sativum*. Am. J. Botany **42**:267–273.

Raju, M. V. S., T. A. Steeves, and J. M. Naylor. 1964. Developmental studies on *Euphorbia esula* L.: Apices of long and short roots. Can. J. Botany **42**:1615–1628.

Reinhard, E. 1954. Beobachtangen an *in vitro* kultivierten Geweben aus dem Vegetationskegel der *Pisum*-Wurzel. Zeit. Botanik. **42**:353–376.

Riopel, J.L. 1966. The distribution of lateral roots in *Musa acuminata* 'Gros Michel.' Am. J. Botany **53**:403–407.

Torrey, J. G. 1952. Effects of light on elongation and branching in pea roots. Plant Physiol. **27**:591–602.

—— 1954. The role of vitamins and micronutrient elements in the nutrition of the apical meristem of pea roots. Plant Physiol. **29**:279–287.

—— 1955. On the determination of vascular patterns during tissue differentiation in excised pea roots. Am. J. Botany **42**:183–198.

—— 1957. Auxin control of vascular pattern formation in regenerating pea root meristems grown *in vitro*. Am. J. Botany **44**:859–870.

THIRTEEN

Differention of

the Plant Body—

Early Stages

The continued growth of the plant body depends upon the production of new cells by mitotic activity in its meristematic regions. One might predict that this would result in a homogeneous cell population, because mitosis ordinarily leads to the formation of identical sister cells. It is obvious, however, that the plant body does not consist of such a uniform assemblage of cells. Rather, it is composed of diverse specialized cells arranged in patterns having functional significance. If this were not the case, the plant could function, at best, in only a very restricted manner. The phenomenon of *differentiation*, as this production of diverse cell types in definite patterns is called, has been alluded to in earlier chapters because it is almost impossible to consider growth apart from it. Now it is necessary to turn attention specifically to the phenomenon itself, one of the major topics of interest in modern developmental biology.

Genetic correlates of differentiation

The diversity of differentiated cell types might suggest that genetic changes must be involved in differentiation. The preponderance of evidence, however, indicates that cellular diversity within the organism is accomplished in a framework of genetic homogeneity and that the genetic changes that do occur are of little significance. The most striking evidence in support of this principle is to be found in the well-known

191

regeneration phenomena characteristic of plants, which will be discussed fully in Chapter Seventeen. Roots, shoots, and in many cases whole plants are often regenerated from fully differentiated tissues either as a normal process or as a result of wounding or some other stimulus. Because the regenerated plants are apparently quite normal and contain all of the cell types characteristic of the species, it must be concluded that the differentiated cells that gave rise to them had not undergone mutation as part of their differentiation. Although it is not possible to observe such regeneration from all tissues of all plants, the phenomenon is of such widespread occurrence that a mutational basis for differentiation is essentially ruled out.

A possible exception to the principle of genetic homogeneity in development has been recognized in the widespread occurrence of quantitative chromosomal changes in differentiated plant cells. Differentiated tissues in many plants have been shown to contain polyploid cells (D'Amato, 1952), but it is equally clear that other plants show none at all (Partanen, 1959). Furthermore, there is no evidence of any regular distribution pattern of polyploid cells that can be correlated with particular tissues or cell types. It is most unlikely, therefore, that these quantitative nuclear changes play a causal role in differentiation. Torrey (1959) has described an interesting situation, however, in a callus derived from *Pisum* roots that may have some bearing upon this question. If the callus is maintained on a complex nutrient medium containing yeast extract and the auxin 2, 4-D, there is a steady increase in the proportion of dividing cells that show various levels of polyploidy and even aneuploidy. On a synthetic medium this change did not occur. Torrey was able to show an inverse relationship between the increase in polyploidy and the ability of the callus to give rise to normal organized roots. On the yeast extract medium, the ability to initiate roots is lost over a period of about six months, and during the period when regeneration can occur the newly formed root meristems are composed entirely of diploid cells. The implication is that in a mixed population of cells only the diploids are able to give rise to organized root meristems. Extrapolating from these observations to the intact plant, one might suggest that quantitative chromosomal changes can impose limitations upon the potency of differentiated cells. Torrey has called attention to the fact that in many roots, even when cells in other tissues become polyploid, the pericycle that normally initiates lateral roots remains diploid. On the other hand, it is not uncommon for polyploid shoots or whole plants to arise by regeneration, so that the limitations of potency seem not always to apply. In reviewing the evidence pertaining to the possible role of quantitative chromosomal changes in differentiation, Partanen has concluded that such changes are incidental to the mechanisms of differentiation and themselves represent a type of cellular differentiation. Whatever the nuclear changes involved in the mechanisms of differentiation, they must be more general in occurrence and must be more directly correlated with specific differentiation.

If differentiation cannot be correlated with mutational or quantitative changes in the genes, it becomes a problem to visualize how so many different cell types can arise under the control of the same genetic material, yet it is unrealistic to seek mechanisms that operate only in the extranuclear portions of the cell. It has long been suggested that the answer to this dilemma must lie in selective gene action so that, though a particular cell type during its development contains all of the genes underlying all of the characteristics of all of the cell types, only certain of these are active in a cell at any one time. Thus there *are* differences among the nuclei of diverse cell types but they are of a functional nature and are not necessarily irreversible.

Only recently has the rapidly developing field of molecular genetics begun to suggest mechanisms by which the selective activation and repression of genes can be accomplished. It has been proposed that other components of the chromosomes, for example histones, may interact with the DNA in such a way that certain genes are repressed, perhaps by physical masking that interferes with the DNA-mediated synthesis of messenger RNA. The mechanism of gene masking is still far from understood, but it does provide a model for selective gene action. The individual activation or repression of genes in development, however, would require a maze of stimuli that would seem to be most unlikely to culminate in organized development in a multicellular organism. This problem has been greatly alleviated by the discovery by Jacob and Monod (1963) that groups of genes are linked together into functional units, each under the control of an operator gene that in turn is controlled by a regulator gene. In this system a single stimulus can result in the activation, or derepression, of all the genes in the operon or functional unit. Thus a relatively small number of stimuli or inducing substances, working through the regulator genes, could set in motion complex developmental phenomena involving many genes. As development continues, new products formed in the cytoplasm can evoke a sequence of new events, and the whole system can remain continuously responsive to external influences. To date this operational mechanism has been elucidated primarily in bacteria, but there is increasing evidence for the existence of similar systems in the higher plants.

This concept of selective gene action seems to provide a reasonable framework for the analysis of mechanisms of cellular differentiation although much further work will be needed to establish its validity in higher organisms. There remains, however, the much more difficult problem of the organized patterns within which cellular differentiation occurs. It is important to know why a particular cell in a developing stem becomes a vessel element; but it is perhaps even more important to understand how this particular cell is integrated into the linear sequence of such elements that constitutes the functional vessel and is in turn part of the vascular system continuous throughout the plant body. Only in this context is the individual vessel element of physiological significance to the plant. The occurrence of levels of organization above the cellular level obviously demands further elaboration of the concepts proposed for cellular phenomena; but it will be well to defer this until after some of the phenomena of differentiation have been considered.

Methods for the study of differentiation

The study of differentiation can begin in a variety of ways. It can begin with mature structures that are traced back to their origins, or with embryonic features that are followed to maturity. It can start with a study of cellular changes later integrated into the overall plan of the organism, or it can explore the origin of the larger plan and then fill in the cellular details. Because the point of view of this book has been a concern for the plan of the organism, it seems appropriate to explore the origin of higher levels of organization first. Because the basic plan emerges in the earliest stages of differentiation, the initial stages of differentiation will be dealt with first in this account, and in Chapter Fourteen the later stages, together with some of the cellular aspects, will be discussed.

It might be expected that the ideal place to begin this consideration of differentiation would be the zygote, and indeed in the earlier discussion of embryology this was done. However, in the plant, developmental phenomena are not concentrated in an embryonic phase early in the life of the organism as is the case in most animals. Rather growth and differentiation are continuing processes associated with the activity of the meristems. Much of the work done on differentiation in plants has been concerned with the derivatives of the meristems rather than the embryo, and particularly with the primary meristems of shoot and root. It is of course recognized that the shoot and root apical meristems themselves reflect a rather high level of differentiation, which becomes apparent when they are set off early in the embryology of the plant as distinct shoot- and root-producing entities. They are, however, stabilized in a state that permits their continued proliferation. On the other hand the derivatives of these meristems undergo further differentiation, which, although it may involve active cell proliferation for a time, ultimately results in a cessation of mitotic activity as maturity is approached. Many cells, of course, if their structural modification does not preclude it, may be reactivated subsequently by a suitable stimulus. With regard to the stabilization of the meristems in a partially differentiated state, it is interesting to note that the vascular cambium apparently represents such a stabilization at a further level of specialization. Thus a consideration of differentiation in relation to the primary meristems does not encompass all aspects of the phenomenon, but it does include most of the work done in this field.

There is a very logical basis for the concentration of effort upon this aspect of differentiation rather than upon embryological stages which should provide closer comparisons with animal studies. The accessibility of the apices of shoot and root for analysis and experimentation has made them attractive objects for investigation. Perhaps more important, however, is the property of repetitive growth that characterizes these regions so that organs, tissues, and cell types are produced predictably in a continuing sequence. This has led to a methodology in plant studies that is not ordinarily encountered in animal embryology. The animal embryologist must study a sequence of individual organisms at different stages of development in order to observe progressive differentiational changes. If a botanist studies plant embryos he must do the same. If however, he investigates a shoot or root apex he may observe stages of differentiation by progressing basipetally from the meristem into the maturing regions of the organ along the axis. He thus effectively substitutes an axis of space for time. Caution must be exercised in the practice of this method because there are ontogenetic changes that disturb the strictly repetitive pattern as in the onset of flowering, and the leaves of the shoot, being determinate organs, of course do not show repetitive growth in their own differentiation. These qualifications, however, pose no problem for the perceptive worker.

Early stages of differentiation in roots

This method of examining the phenomena of differentiation is equally applicable to shoot and root apices, but because of the presence of leaves and the influences they exert, the patterns in the shoot are more complex. Consequently it seems appropriate to begin this consideration with the root. Differentiation in the root of *Pisum sativum*, the garden pea, has been described in some detail by Torrey (1955) and by Popham (1955) and in addition it has been a useful object of experimental analysis of various aspects of the process. The root promeristem of *Pisum* does not consist of superposed meristematic layers which could be interpreted as histogens so that differentiation of tissues may be considered without this complication (Fig. 12.3). Immediately proximal to the promeristem of the root may be detected the first indications of tissue differentiation, but as might be expected they are vague and rather generalized (Fig. 13.1a). This initial differentiation leads to the establishment of three tissue systems: dermal, fundamental or ground, and vascular, distinguishable primarily because of the varying planes of cell division and the early vacuolation of ground tissues. In the ground tissues of the future cortex early divisions are predominantly transversely oriented. This leads to the formation of files of cells extending back along the root axis (Fig. 12.3). However, the number of files of cells in the cortex increases basipetally as a result of periclinal divisions in the innermost layer, and in transverse sections a highly regular radial arrangement of cells may often be seen.

The vascular system of the root is a solid core of xylem, fluted in outline, with phloem peripheral to it contained in the bays of the xylem. The protostelic nature of the future vascular system is evident immediately behind the meristem where it is recognizable as a solid core of rather small densely staining cells within the vacuolating ground meristem of the cortex (Fig. 13.1b). In these cells the plane of division is predominantly longitudinal with the result that cell length soon exceeds transverse width. When the cells have achieved this form it is customary to designate the tissue as *procambium*, a generally recognized early stage in vascular differentiation. Separating the procambium from the surround-

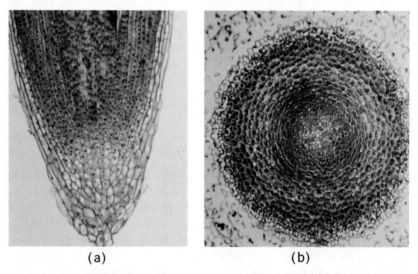

(a)   (b)

Fig. 13.1   Initial differentiation of tissues behind the root apex of *Pisum sativum* (pea). (a) Longitudinal section of the root tip. The central vascular tissue consists of narrow elongated cells, and the superficial dermal tissue consists of cubical cells. Between the two is the ground tissue in which cell shape is cubical. The root cap covers the root surface at this level. (b) Transverse section just behind the terminal meristem. The central vascular tissue consists of very narrow cells. In the vascular tissue the centrally located procambium that will differentiate as metaxylem has started to vacuolate. ×58.

ing ground meristem is a single layer of cells, the *pericycle*, which becomes distinct at a very early stage and which is of significance as the later source of lateral roots.

These considerations of initial differentiation lead to the question of the distribution of cell division and cell enlargement in the root tip. Any suggestion that division, enlargement, and differentiation are spatially separated along the axis must obviously be discarded. As pointed out in Chapter Twelve, divisions in the root are not limited to the region of the meristem but occur also in the differentiating derivative cells. Several workers have devised methods for determining the frequency of cell division along the axis of the root. Goodwin and Avers (1956) grew seedlings of *Phleum* in chambers that permitted continuous observation and time lapse photography of superficial cells of the growing zone of the root. They have reported that the frequency of cell division is low in the region of the meristem, increases to a maximum at about 150 to 200 microns behind the root cap, and declines to zero at about 400 microns. Net cell elongation begins as mitotic frequency declines and the epidermal cells attain their final length at slightly more than a millimeter from the tip. Erickson and Sax (1956a, b) obtained information on cell division and elongation in the root epidermis of *Zea* and their data confirm these findings. They do, however, show marked quantitative differences between these two grass species. The location of maximum frequency of cell division in *Zea* is 1.25 millimeters behind the tip of the meristem and divisions continue to a total distance of two millimeters from the tip. Net cell elongation begins with the same relationship to cell division as in *Phleum* but continues to a distance of 9.5 millimeters from the tip. These data indicate the overlapping of cell division and cell elongation in differentiating cells, but because they apply to the epidermis only, it is difficult to draw conclusions applicable to other tissues of the root.

The analysis of cell division frequency and distribution in root tips was extended by Jensen and Kavaljian (1958) to include all of the tissues in the root of *Allium cepa*. In order to study internal tissues, however, it was necessary to work with prepared root sections so that the advantage of continuous observations on living cells was lost. As in the earlier studies, these investigators found that the frequency of mitosis in the meristem region is low (Fig. 13.2). In fact, they considered that their data supported the concept of the quiescent center that had been proposed two years earlier by Clowes on entirely different grounds. The frequency of cell division rose and then declined to zero basipetally along the axis, but the most significant finding was that the peak of mitotic frequency occurred at a different position in each of the three tissue systems, being reached first in the ground tissue of the future cortex, then in the vascular core, and finally at about a millimeter from the tip

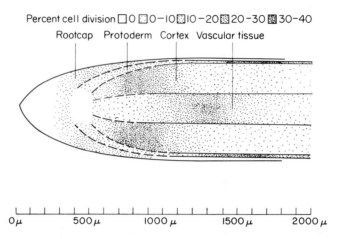

Percent cell division ☐ 0 ▤ 0–10 ▨ 10–20 ▧ 20–30 ▦ 30–40

Fig. 13.2   The distribution of cell division in the onion root tip. For each tissue the highest rate of cell division occurs at a different distance behind the tip. The data shown are pooled data from five root tips, and were measurements made on noon collections. It was found that the percent of dividing cells varied at different times of the day, but the noon values were high for all tissues. (W. A. Jensen and L. G. Kavaljian, 1958. Amer. J. Bot. **45**: 365.)

of the meristem, in the dermal tissue. There is also a difference among the tissues in the division frequency achieved at the peaks. Thus it is apparent that differences in the frequency of cell division form an important part of the initial differentiation of tissues in the root.

Early stages of differentiation in shoots

Histodifferentiation in the shoot can be studied in the same manner employed for the root. However, the pattern of differentiation is so markedly influenced by the developing leaves that even a morphological description of developmental phenomena in the stem is much more complex than in the root. It cannot be doubted that the physiological basis of differentiation is equivalently complex. In Chapter Ten the segmental nature of the stem with its nodal-internodal organization was discussed in relation to the question of elongation. This pattern has its counterpart in the internal organization of the stem, which must be considered in any treatment of tissue differentiation.

Topography of the mature
vascular system

Although our concern is the pattern of differentiation of the principal tissues of the shoot, it is appropriate to begin by describing the structure of the mature organ. This is because the course traced by the vascular tissue system in particular is so complex that its differentiation can be comprehended best after the final state is understood.

Examination of the course of the mature vascular system in the stem of a plant such as *Coleus, Impatiens,* or *Lupinus*—by the study either of serial transverse sections or of stems cleared by chemical treatment for translucence that enables the vascular tissue to be seen internally—reveals that it consists of a system of discrete strands called vascular bundles (Fig. 13.3a). The bundles, each of which consists of both xylem and phloem, extend along the axis as an interconnected system which forms a ring around the central pith. The system may be regarded as anastomosing if one begins the examination near the apex of the stem and proceeds basipetally; but it would be described as a system of branching bundles if examined from the base toward the tip. The most striking feature of this vascular system is that at each node one or more bundles turn out

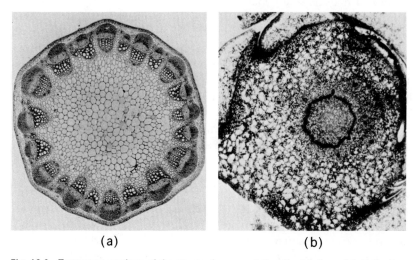

(a)　　　　　　　　　(b)

Fig. 13.3 Transverse sections of the stems of representative dicotyledons. (a) *Helianthus annuus* (sunflower). The vascular tissue is arranged in a series of separate bundles. (b) *Rhododendron grandifolia*. The vascular tissue forms a complete cylinder that is interrupted only by the very small leaf gaps associated with the leaf trace attachment. (a) ×8, (b) ×2.

from the central cylinder and extend across the cortex into the petiole of the leaf borne at that node. If a vascular bundle extending through a leaf petiole as a leaf trace is followed basipetally into the central cylinder it may be found to join directly to another bundle which did not diverge at that node, or more commonly it may be seen to extend downward, sometimes vertically, sometimes obliquely, through one or more internodes before joining another bundle, and other smaller bundles may join it in its basipetal course.

Thus the impression emerges of a unitary vascular system in stem and leaves that could be separated into cauline and foliar components only with great difficulty if at all. In fact, the vascular system in the stem may be described as being largely, if not entirely, leaf-oriented. This orientation seems to be eminently sound physiologically, considering that the vascular system is the conducting apparatus of the plant and that the leaf is the major source of elaborated organic compounds and the destination of most of the water that passes through the stem. This vascular pattern may be said to be characteristic of the seed plants generally and it contrasts sharply with those found in the lower vascular plants. There are, however, variations that must be recognized. In some plants such as *Syringa*, *Dianthus* and *Hypericum*, the bundles in the stem are extremely broad and may be contiguous, so that in transverse section the vascular cylinder seems to be in the form of a continuous ring of xylem and phloem (Fig. 13.3b).

In those species in which the primary vascular tissue of the stem forms an essentially continuous ring it is possible to observe the features known as *leaf gaps* with great ease. Where a vascular bundle diverges into a leaf petiole, the region of the vascular ring that confronts it, through which it would have extended had it not diverged, is occupied by parenchyma rather than by xylem and phloem. Ordinarily the vascular tissue on either side of the gap closes over it at some distance above the departure of the leaf trace so that the leaf gap bears some resemblance to a Gothic window, with the limitation, of course, that the gap is not an empty space but is filled with parenchyma. Thus the pith and the cortex are continuous through the leaf gap. If the vascular system is composed of widely spaced bundles, leaf gaps are more difficult to visualize; but it is evident that the parenchymatous region that confronts a departing leaf trace corresponds exactly to a typical leaf gap and should be considered as one. Frequently such a gap is closed above by the joining of bundles lateral to it, but this may occur only at a great distance above the departing trace, or it may not occur at all. A difficulty is that not all of the parenchymatous intervals in the vascular ring can be considered to be leaf gaps, whereas this is the case when the vascular tissue forms a continuous ring.

Where vascular patterns in the stem have been analyzed three-dimensionally it has been found that the bundles show a high degree of regularity in their sequence of interconnection. Not surprisingly, these patterns can be correlated with the phyllotaxy of the shoot, and a knowledge of the phyllotaxy can be most useful in interpreting a particular system. If each leaf is served by a single trace the pattern may be relatively simple; but where multiple traces occur, each with its own course through the stem and ultimate connection with other bundles, the system may be extremely complex. This is very well illustrated in many monocotyledons where large numbers of traces enter the sheathing bases of the leaves. To the student of developmental morphology the close correlation between the stem vascular system and the leaves provides clear evidence of a developmental relationship, and it is to the exploration of this relationship that attention now must be turned.

Differentiation related to leaves

The study of tissue differentiation in the stem ordinarily is carried out by examination of both transverse and longitudinal serial sections of the shoot apex together with the subjacent regions in which maturation is occurring. This is essential because the often oblique course of bundles may lead to erroneous interpretations of their course and connections if they are seen only in longitudinal sections. This method and the kind of result it yields are illustrated in a study of tissue differentiation in the shoot of *Linum perenne* carried out by Esau (1942). The terminal meristem is a high, domed mound, and leaf primordia are initiated on its sloping flanks (Fig. 13.4). A distinctive feature of differentiation in the stem immediately behind the meristem is that there are differences among the tissues in the rapidity of maturation. Cells of the pith begin vacuolation very close to the summit of the apex and enlarge much more rapidly than do the cells that surround them peripherally. Moreover at the surface of the apical flanks the protoderm forms a distinctive superficial layer. Between these two distinctive regions are the relatively unvacuolated cells that will give rise to the cortex and the vascular tissue. It is from this peripheral region also that the leaves arise. The close relationship between the leaves and the vascular tissue of the stem can be seen to have its origin at this level for, from the time of its inception, each leaf primordium has associated with it and extending into it a recognizable strand of procambium (Fig. 13.5a). At its uppermost limits the procambium is identifiable both by the shape of its constituent cells and by their intensified staining reaction to a variety of histological stains. Procambium is initiated by longitudinal divisions that occur with greater frequency than in surrounding cells with the result that procambial cells have an elongate

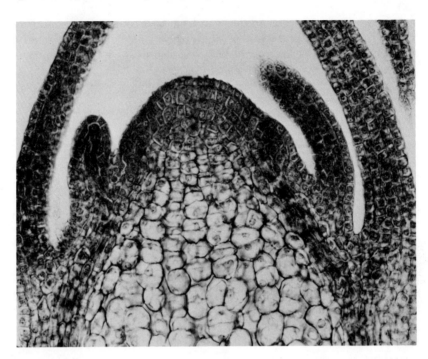

Fig. 13.4   A median longitudinal section of the shoot apex of *Linum* showing early differentiation of the pith and initiation of the procambium in relation to a young leaf primordium on the right side of the meristem. ×400.

shape, that is, they are longer than they are wide. However, they are initially not necessarily longer than are surrounding cells, although this is clearly true at later stages of development. Thus the developmental phenomenon that initiates procambium is a preferential orientation of the planes of cell division in localized groups of cells. This does not mean that procambial initiation occurs in isolation. Rather, it is clear that this process represents the upper extremity of a sequence of similar divisions occurring in a continuous column of differentiating vascular tissue. Thus the procambial system differentiates acropetally into the growing apex and in continuity with the existing vascular system. Reports of discontinuity or of basipetal procambial differentiation in shoot apices appear to have been in error and to have resulted from a failure to recognize the oblique course that the procambial strand may follow, causing them to appear to end blindly when observed in longitudinal sections (Fig. 13.5a, b). There are, however, some circumstances in which procambium does differentiate basipetally and examples of these will be discussed in their proper context.

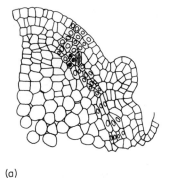

(a)

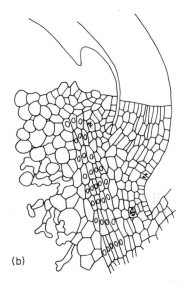

(b)

Fig. 13.5 Longitudinal sections of the shoot apex of *Linum perenne* showing the correlated initiation of a leaf primordium and its associated procambial strand. Cells with nuclei
drawn in are those involved in leaf and procambium development. (a) The upper part of the
procambial strand showing the association with the leaf primordium at the time of inception.
(b) The lower part of the same strand seen in an adjacent longitudinal section where the
strand has diverged around the gap of an older leaf. ×300. (K. Esau, 1942. Amer. J. Bot.
29 : 738.)

In the procambial system the pattern of the primary vascular tissue of
the shoot is established. In the terminal region of the shoot it is much
foreshortened as is the stem itself, but it becomes extended in the process
of internodal elongation. This is particularly conspicuous in the development of long shoots. During this process there is considerable net cell
elongation in the procambium, in contrast to the ground tissues where
there is a preponderance of transversely oriented cell divisions. In a
series of transverse sections progressing basipetally from the level of
initiation, it can be seen that the number of procambial cells in each
bundle increases (Fig. 13.6). This is accomplished by longitudinal divisions in the procambial cells themselves and by addition of cells from
surrounding regions by the appropriate orientation of cell division planes.
This process is of particular importance in those species in which the
bundles become very wide and in effect form a continuous primary
vascular cylinder.

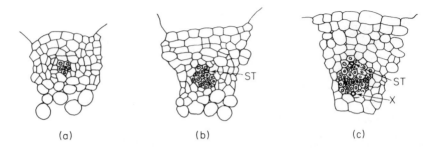

(a)   (b)   (c)

Fig. 13.6 Diagrams of procambial strands associated with successively older leaves of *Linum perenne*. The strands are seen in transverse section. Procambial cells have the nuclei indicated, and it can be seen that the number of cells in the strands increases ontogenetically. Key: ST, sieve tube; X, xylem. ×300. (K. Esau, 1942. Amer. J. Bot. **29**: 738.)

Differentiation related to the terminal meristem

Before considering the further differentiation of procambium and the final maturation of procambial cells as xylem and phloem, a question concerning initial differentiation must be discussed. This is the question of whether, in fact, the procambium differentiated in relation to leaf primordia is the initial stage of vascular tissue differentiation or whether there is a stage in the process that precedes procambium. Because the evidence pertaining to this question must be obtained in the region immediately subjacent to the terminal meristem where cellular characteristics are at best indistinct, it is not surprising that the question has not been settled. Nonetheless it is an important issue because it pertains to the interpretation of the relative influences of terminal meristem and leaf primordia in regulating vascular differentiation. If the procambium, initiated in strands associated with the leaves, is the first stage, it might be argued that the terminal meristem plays a passive role in the process and the developing leaves determine the pattern of the vascular system. If, on the other hand, there is a stage preceding procambial initiation, then the suggestion might be that this earlier stage, influenced by the terminal meristem, sets out the basic plan of the vascular system, a plan that is subsequently modified by the influence of the developing leaves. Although final answers are not yet available, it may be well to examine the evidence pertaining to this difficult question as an indication of the current status of investigations.

In a recent description of vascular differentiation in *Geum chiloense*, McArthur (1967) examined the relationship between leaf primordia and the earliest sign of vascular differentiation. This study confirmed the relationship between leaf primordia and procambial initiation. However, it was found that the procambium strands were initiated within a cylinder of small, densely staining cells that resemble procambial cells in cytological characteristics but not in size or shape (Fig. 13.7).

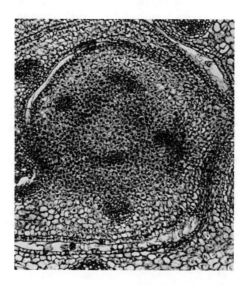

Fig. 13.7 Transverse section 80 microns below the apex of the stem of *Geum chiloense*. There is a distinct ring of provascular tissue within which procambial strands are differentiating in relation to developing leaf primordia. ×90. (Photograph courtesy I. C. S. McArthur.)

Because the leaf-oriented procambial strands arose as part of this cylinder only, it does not seem likely that the cylinder itself can be interpreted as being related to the leaves in its differentiation. Rather it seems to arise under the influence of the terminal meristem. This cylinder appears to be the forerunner of the vascular system of the stem and as such may be designated *provascular tissue*. Although no procambium could be detected in relation to the youngest leaf primordium, it was present in all others and was recognizable at essentially as high a level in the stem tip as any other part of the cylinder. The heterogeneity of the cylinder at its first appearance results from differences in timing in the early differentiation of procambium. In a sector of the apex occupied by a young leaf primordium, underlying provascular tissue is converted almost immediately to procambium. In interfoliar positions at the stem tip the cylinder consists of typical provascular tissue, some of which will later be converted to procambium in the acropetal differentiation of strands which will form in relation to primordia initiated later.

The concept of a stage of vascular differentiation preceding procambial initiation is not a new one and the tissue here described as provascular tissue has appeared repeatedly in the literature under a variety of names including *prodesmogen*, *meristem ring*, *Restmeristem* or *residual meristem*, and *prestelar tissue*. There are two rather different interpretations of the tissue described by these various terms and the difference between the two is not merely one of terminology. On the one hand, the faintly delimited cylinder of meristematic tissue in the axis immediately subjacent to the terminal meristem is interpreted, as in the description of *Geum*, as the first stage of vascular differentiation occurring more or less independently

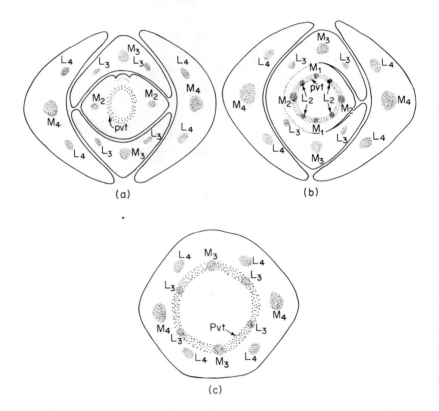

(a)

(b)

(c)

Fig. 13.8 The initiation of vascular tissues in the stem of *Garrya elliptica* seen in a series of transverse sections at increasing distances behind the apex. (a) Section in the internode below leaf pair 2. In the stem the vascular tissue is present as a ring of provascular tissue. (b) Section at the level of the third node. The vascular tissue in the stem consists of provascular tissue part of which has differentiated as median and lateral trace procambium associated with the second leaf pair above. (c) A section lower in the stem showing more of the provascular tissue differentiated as procambium. Key: L, lateral leaf trace; M, median leaf trace; PvT, provascular tissue. (R. M. Reeve, 1942. Amer. J. Bot. **29**: 697.)

of the leaves, but in relation to the activity of the terminal meristem (Fig. 13.8). Further differentiation to the procambial stage occurs only in relation to the leaves; but the provascular tissue, as it may appropriately be designated in this context, is held to represent a differentiational departure from the promeristem condition in the direction of vascular tissue. On the other hand, another view holds that the tissue in question has not undergone the first stages in vascular differentiation but rather represents a perpetuation of the uncommitted meristematic state and is recognizable only because of the precocious differentiation of the ground tissue of pith and cortex. In this view the procambium does represent the first step in differentiation toward vascular tissue, and the tissue of the cylinder consists of *residual meristem*—that is, cells in a still uncommitted state—which may or may not become procambium depending upon the relation to developing leaf primordia. It is not difficult to understand why this difference of interpretation remains unresolved.

Although there is no certainty that vascular differentiation is identical in all groups of plants, it is not unreasonable to think that a study of vascular cyptogams might throw some light upon the process in seed plants. In fact, several studies of vascular differentiation in these plants have been carried out. In *Lycopodium* Freeberg and Wetmore (1967) have described a central column of procambium in the stem tip extending above the level of the youngest leaf primordia, which are themselves initially without procambium. In this protostelic plant the leaves appear to exert no influence upon the initiation of the stem vascular system so that in many respects the differentiation pattern is like that of the root. In the ferns, because the terminal meristem consists of enlarged, highly vacuolated cells it may be possible to examine initial differentiation of vascular tissues without the complication of the probable resemblance of the earliest stages to the cells of the terminal meristem. This expectation was realized in *Osmunda cinnamomea* in which Steeves (1963) recognized, almost immediately beneath the enlarged cells of the surface prismatic layer, cells appearing to be in the early stages of differentiation and quite different in cytological characteristics from the prismatic cells (Fig. 13.9a). In the center was a cluster of isodiametric cells that, because they appeared to be ontogenetically related to the pith rib meristem and the pith, were considered to be *pith mother cells*. Around these in the form of a flattened, truncated cone was a layer of smaller, more intensely staining cells. Because these were in basal continuity with obvious procambium they were termed collectively *incipient vascular tissue*. This tissue appears to be comparable to provascular tissue of other stems. In view of the distinctive characteristics of these cells in comparison with those of the prismatic layer in *Osmunda*, it would be difficult to consider them as constituting a mere residual meristem. In the youngest leaf primordia,

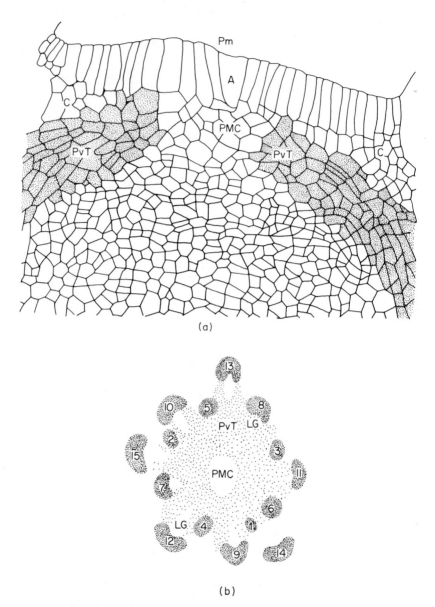

(a)

(b)

Fig. 13.9 Interpretation of initial vascular differentiation in the shoot of *Osmunda cinnamomea*. (a) Longitudinal section of the shoot apex showing the relation of the provascular tissue to the terminal meristem. (Compare with Fig. 4.10.) (b) Transverse section 170 microns behind the promeristem showing provascular tissue, differentiating leaf trace procambium, and the origin of leaf gaps. Key: A, apical cell; C, cortex; PvT, provascular tissue; LG, leaf gap; Pm, promeristem; PMC, pith mother cells; 1–15, leaf trace procambium of leaves 1–15. (Adapted from T. A. Steeves, 1963. J. Indian Bot. Soc. **42A**: 225.)

incipient vascular tissue, similar to that of the stem and in continuity with it, could be detected. Procambium was differentiated acropetally into the leaf primordia within the incipient vascular tissue and in continuity with older procambium at lower levels in the stem. Above the level of the fifth or sixth youngest leaf primordium, the layer of incipient vascular tissue in the stem was uninterrupted; but in relation to the older leaf primordia, parenchymatous gaps began to develop confronting the procambial leaf traces (Fig. 13.9b). Thus the originally continuous cone of incipient vascular tissue was interrupted by the development of leaf gaps. Thus the concept of initial differentiation of the vascular system independent of the leaves but subsequently modified in relation to leaf development is supported by the pattern of differentiation in *Osmunda* in which, as in the seed plants, procambium appears to be initiated largely if not entirely in relation to leaf primordia.

A similar interpretation of the initial differentiation of vascular tissues in the fern shoot apex had been advanced earlier by Wardlaw (1944a) based upon observations of *Dryopteris* and several other species. In his descriptions, however, the incipient vascular tissue was considered to constitute a solid core beneath the prismatic layer and the pith was held to arise by the further changes in cells of potentially vascular tissue. The difference between this interpretation and that derived from the study of *Osmunda* may be reduced to the question of whether the centermost cells in the region of initial differentiation are different from the peripheral cells from the outset or whether they become different only on further divergent differentiation. It must be recognized that there may well be differences among species in this regard, and both interpretations are in agreement as to the initial delimitation of the vascular system independently of the leaves and its subsequent modification in relation to leaf development.

On the other hand, several workers also examining the early stages of differentiation in ferns came to different conclusions. Kaplan (1937), who proposed the *Restmeristem* or *residual meristem* interpretation in the seed plants, was able to account for the histological features that he observed in fern shoot apices within the same framework; and Hagemann (1964) failed to detect anything that he could call potentially vascular tissue in the shoot apex above the level of procambium in the leaf traces. Thus histology alone cannot be relied upon to resolve the question of the identity of the first stage in vascular differentiation. Fortunately there is a substantial body of experimental evidence for ferns that deals with this question and lends support to the existence of a definite incipient vascular or provascular stage preceding the procambial stage in the shoot apex.

Experimental analysis of early stages
of vascular differentiation

In *Dryopteris* and several other ferns, Wardlaw (1944b, 1946) sought to eliminate the influence of developing leaf primordia on vascular development by puncturing successive leaves at the time of, or shortly after, their inception. He reasoned that such an operation ought to reveal the extent to which the terminal meristem alone is capable of initiating a vascular system, and the nature of the modifications that the developing leaves exert upon that system. The result of this experiment in several species was the differentiation of a complete ring of mature vascular tissue uninterrupted by leaf gaps (Fig. 13.10). Thus in the absence of

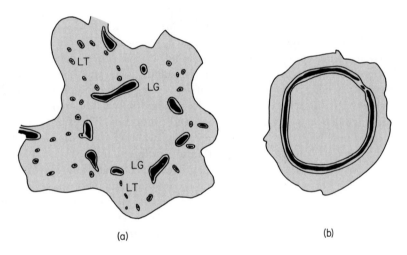

(a)   (b)

Fig. 13.10   The effect on differentiation of vascular tissue in the stem of *Dryopteris dilatata* of systematically puncturing leaf primordia at the time of their initiation. (a) Distribution of vascular tissue in the intact stem seen in transverse section. (b) Transverse section of a stem in which successive leaf primordia had been punctured. The vascular tissue forms an uninterrupted ring. Key: LG, leaf gap; LT, leaf trace. ×3. (C. W. Wardlaw, 1944. Ann. Bot. **8**: 387.)

leaves the terminal meristem does produce a vascular system, thereby providing indirect evidence for the validity of the interpretation of a provascular or incipient vascular cone differentiated in the apex independently of leaves. The experiment also reveals the modifying influence of leaves on the initial vascular system because it appears that they are causal in the differentiation of leaf gaps.

These experiments have been criticized on the grounds that the conclusions were based upon the configuration of the resulting mature stem tissues and did not consider the actual development of the experimental plants. Although it is most unlikely that the mature structures described could have developed in any way other than that suggested, it is reassuring that more recent experiments of a comparable nature which included a developmental analysis have led to the same conclusion. Soe (1959) punctured young leaf primordia in the fern *Onoclea sensibilis* as they were formed and found that, although the resulting vascular system had very small leaf gaps, they were nevertheless present. However, when the sites of prospective leaf primordia were punctured before the leaf actually protruded from the surface of the meristem, the resulting stem vascular system formed an uninterrupted cylinder of xylem and phloem. In the unoperated apex of *Onoclea*, although procambium initiation in relation to leaf primordia is readily observed, it is difficult to detect incipient vascular tissue as a continuous cylinder. In experimentally treated apices, however, such an uninterrupted cylinder is clearly present and is basally continuous with procambium. Comparison of the two kinds of apices reveals that the difficulty in observing incipient vascular tissue in the normal apex is the result of the very early development of large leaf gaps in relation to the youngest primordia. This also explains why puncturing leaf primordia after they have emerged does not prevent gap formation, the gap already being present at this stage of development. Thus the experimental study of differentiation processes in apices lends strong support to the conclusions reached from earlier studies.

In view of the usefulness of the experimental approach in elucidating initial vascular differentiation in ferns, it would seem reasonable to apply the same technique to flowering plants. In fact, some experiments of this type have been done, but the results have given answers that are less clear than in the ferns. Wardlaw (1950) and Ball (1952) carried out surgical experiments that had the effect of removing, or at least greatly reducing, leaf influence upon the stem vascular system during its development in the apices of *Primula* and *Lupinus*, respectively. In both cases the terminal meristem was first isolated laterally by vertical incisions. During the ensuing period of reorganization the shoot meristem developed for a time without forming leaf primordia. In the case of *Primula* subsequently formed leaf primordia were punctured as they emerged for a time, but ultimately, in both cases, leaves were allowed to develop on the isolated meristem. Under these conditions in which the leaf influence was reduced, or probably absent for a time, the development of a provascular cylinder was more readily detectable than in unoperated apices; but because leaves were subsequently permitted to develop it is difficult to determine the extent of the leaf influence. The fact that the resulting mature vascular system tended to be in the form of separate bundles suggests that the leaf influence may have been rather large. This last conclusion is supported by the work of Young (1954), also on *Lupinus*, which showed that if all leaves were systematically removed at an early primordial stage a distinct provascular cylinder developed but there was no further differentiation of this tissue into procambium or mature vascular tissue.

Some of the confusion surrounding the question of initial vascular differentiation in flowering plants has been resolved in a study by McArthur (1967), referred to earlier in this chapter, which was designed to examine differentiation of vascular tissues in the stem under conditions where the influence of the leaf could be evaluated accurately. The terminal meristem of *Geum chiloense* shoots was isolated laterally by three vertical incisions and successive leaf primordia were punctured as they appeared. When these apices were fixed and examined histologically they were found to contain a provascular cylinder throughout the region of new growth. The provascular cylinder was also prolonged basally as an indistinct zone into the supporting pith plug, presumably by the conversion of previously existing immature pith cells. Thus, a provascular ring was produced and maintained in the stem in the absence of leaves as in Young's experiment on *Lupinus*. It is important that no further development of provascular tissue, even to the procambial stage, occurred in the absence of leaves.

In further experiments, after several leaves had been punctured on the partially isolated meristem, later leaves were allowed to develop for several weeks. These shoots contained a vascular pattern rather like that of the normal shoot with procambial traces containing some mature vascular tissue associated with the developing leaves. The procambial strands had also extended basipetally into the supporting pith plug. This suggests that the leaf is essential for further development of the provascular tissue. Because developing leaves are known to be important sources of auxin in the shoot apex, their influence was replaced in the leafless experimental apices by applying the auxin indoleacetic acid in lanolin as a cap over the apex. Following this treatment the provascular ring became much more distinct and easily recognizable, but there was no differentiation beyond this stage (Fig. 13.11a). Thus, auxin alone does not replace the influence of the leaf primordium in promoting vascular differentiation beyond the provascular stage. From other observations it seemed that carbohydrate nutrition might be a limiting factor. Auxin treatment was therefore combined with application of a sucrose solution in which the bases of the shoots were immersed. Growth of the isolated apices was considerably enhanced and more complete vascular differentiation occurred. Apart from the terminal portion of the isolated apex, which contained a distinct provascular core or ring, all of the other vascular

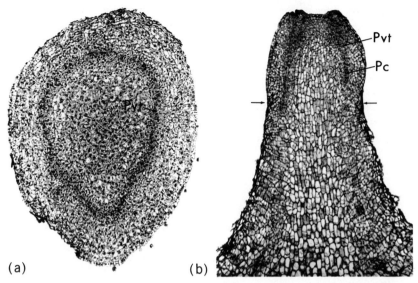

(a)   (b)

Fig. 13.11 (a) Transverse section 270 microns below the tip of a surgically isolated apex of *Geum chiloense* on which successive leaf primordia had been punctured at the time of their initiation and to which indoleacetic acid had been applied. Note the distinct ring of provascular tissue that has been formed. (b) Longitudinal section of an isolated shoot apex of *Geum* after leaf puncturing, application of indoleacetic acid, and addition of sucrose. Provascular tissue in the apical region has differentiated as procambium at lower levels in the stem. Arrows indicate the junction between the newly formed shoot above and the supporting pith plug below. Key: Pc, procambium; PvT, provascular tissue. (a) ×60, (b) ×70. (Photographs courtesy I. C. S. McArthur.)

tissue of the isolated shoot had differentiated at least to the procambial stage (Fig. 13.11b). A procambial ring without leaf traces extended through most of the experimentally produced shoot. Xylem elements were present in the pith plug below the isolated shoot, but it is not known whether final vascular maturation would have occurred throughout the procambial ring if the experiment had been continued for a longer time. However, the formation of procambium indicates the potentiality of the provascular tissue to differentiate as a vascular cylinder in the absence of leaves if suitable nutritional and hormonal factors are provided artificially.

It now seems possible to suggest a resolution of the apparent disparity between the descriptions of initial differentiation of vascular tissues in ferns and seed plants. In both it appears that the shoot apex directly promotes an initial blocking out of the vascular system in the form of a provascular ring. In the ferns the vascular differentiation seems to proceed to completion without the contribution of leaves, although the leaves normally produce extensive modifications in the pattern of development. In the seed plants, on the other hand, later stages of differentiation, including the procambial stage, seem to be dependent upon the contribution of leaves and in their absence do not occur. The leaf effect can be artificially replaced, but its exact nature is not clear at the present time, although auxin would appear to be implicated. The experiments reported here point out a potentially fruitful approach to the analysis of relative contributions of the terminal meristem and leaf primordia to the complex process of vascular differentiation in the shoots of vascular plants.

General comment

The view that emerges from this examination of initial stages of differentiation in roots and shoots is one of blocking out of major tissue regions in relation to the activities of the terminal meristems. These meristems, which in Chapters Six and Twelve were shown to be autonomous with respect to their ability to develop normally when isolated from the rest of the plant, are now seen also to be the organizers of development in the organs to which they give rise. Thus in a real sense it can be said that meristems make plants.

On the other hand, the control that the meristems appear to exercise over the differentiation of their cellular products may be somewhat deceptive in that they regulate only the continued differentiation of tissues that had their origin in the embryo. It will be recalled that the initial differentiation of procambium, protoderm, and ground tissue in the embryos described in Chapter Two often preceded the establishment of morphologically distinct, organized meristems at the two poles of the embryonic axis. It is not known whether the initial differentiation of these tissues in the embryo is regulated by the meristems, and this leads to the conclusion that the control of morphogenetic processes at the time of their inception may be different from the regulation of their perpetuation in later stages of development. At present there is little evidence to support the suggestion that different controls operate in the blocking out of tissues at different stages of plant growth, but it is a view that must be taken into consideration in any comprehensive treatment of differentiation in the plant body.

REFERENCES

Ball, E. 1952. Morphogenesis of shoots after isolation of the shoot apex of *Lupinus albus*. Am. J. Botany **39**:167–191.

D'Amato, F. 1952. Polyploidy in the differentiation and function of tissues and cells in plants. Caryologia **4**:311–358.

Erickson, R. O., and K. B. Sax. 1956a. Elemental growth rate of the primary root of *Zea mays*. Proc. Am. Phil. Soc. **100**:487–498.

———— 1956b. Rates of cell division and cell elongation in the growth of the primary root of *Zea mays*. Proc. Am. Phil. Soc. **100**:499–514.

Esau, K. 1942. Vascular differentiation in the vegetative shoot of *Linum*. I. The procambium. Am. J. Botany **29**:738–747.

Freeberg, J. A., and R. H. Wetmore. 1967. The Lycopsida—A study in development. Phytomorphology **17**:78–91.

Goodwin, R. H., and C. J. Avers. 1956. Studies on roots. III. An analysis of root growth in *Phleum pratense* using photomicrographic records. Am. J. Botany **43**:479–487.

Hagemann, W. 1964. Vergleichende Untersuchungen zur Entwicklungsgeschichte des Farnsprosses. I. Morphogenese und Histogenese am Sprossscheitel leptosporangiater Farne. Beit. Biol. Pflanzen. **40**:27–64.

Jacob, F., and J. Monod. 1963. Genetic repression, allosteric inhibition, and cellular differentiation. *In* M. Locke [ed.], Cytodifferentiation and macromolecular synthesis. 21st Growth Symp., pp. 30–64. Academic Press, New York.

Jensen, W. A., and L. G. Kavaljian. 1958. An analysis of cell morphology and the periodicity of division in the root tip of *Allium cepa*. Am. J. Botany. **45**:365–372.

Kaplan, R. 1937. Uber die Bildung der Stele aus dem Urmeristem von Pteridophyten und Spermatophyten. Planta. **27**:224–268.

McArthur, I. C. S. 1967. Experimental vascular differentiation in *Geum chiloense* Balbis. Thesis, University of Saskatchewan.

Partanen, C. R. 1959. Quantitative chromosomal changes and differentiation in plants. *In* D. Rudnick [ed.], Developmental cytology. 16th Growth Symp. pp. 21–45. Ronald, New York.

Popham, R. A. 1955. Levels of tissue differentiation in primary roots of *Pisum sativum*. Am J. Botany **42**:529–540.

Soe, K. 1959. Morphogenetic studies on *Onoclea Sensibilis* L. Thesis, Harvard University.

Steeves, T. A. 1963. Morphogenetic studies on *Osmunda cinnamomea* L.: The shoot apex. J. Indian Botan. Soc. **42A**:225–236.

Torrey, J. G. 1955. On the determination of vascular patterns during tissue differentiation in excised pea roots. Am. J. Botany **42**:183–198.

———— 1959. Experimental modification of development in the root. *In* D. Rudnick [ed.], Cell, organism and milieu. 17th Growth Symp. pp. 189–222. Ronald, New York.

Wardlaw, C. W. 1944a. Experimental and analytical studies of pteridophytes. III. Stelar morphology: The initial differentiation of vascular tissue. Ann. Botany (London) (N. S.) **8**:173–188.

Wardlaw, C. W. 1944b. Experimental and analytical studies of pteridophytes. IV. Stelar morphology: Experimental observations on the relation between leaf development and stelar morphology in species of *Dryopteris* and *Onoclea*. Ann. Botany (London) (N. S.) **8**:387–399.

———— 1946. Experimental and analytical studies of pteridophytes. VII. Stelar morphology: The effect of defoliation on the stele of *Osumunda* and *Todea*. Ann. Botany (London) (N. S.) **10**:97–107.

———— 1950. The comparative investigation of apices of vascular plants by experimental methods. Phil. Trans. Roy. Soc. London Ser. **B234**:583–604.

Young, B. S. 1954. The effects of leaf primordia on differentiation in the stem. New Phytologist **53**:445–460.

FOURTEEN

Differention of

the Plant Body—

Later Stages

The early stages of differentiation, considered in Chapter Thirteen, are characterized by the blocking out of regions within which cells are relatively homogeneous and behave similarly. In contrast, the later stages of differentiation show highly localized specializations, often with adjacent cells differing markedly in the developmental changes that they undergo. Therefore, in studying these stages of differentiation it is essential to pay close attention to the events taking place in individual cells; and this will be the first task. When this has been done it will be necessary to return to the higher levels of organization to consider the interrelations that keep the differentiating cells as part of an organized system.

Cellular changes during differentiation

In contrast to the body of higher animals with its large number of differentiated cell types, relatively few cell types, possibly no more than 12, are differentiated in the body of the vascular plants. However, to describe the cellular events related to differentiation of even this small number of cell types would fill many pages of this book. Rather than do this, attention will be concentrated on the differentiation of the tracheary elements of the xylem, the tracheids, and vessel members. These are functionally important in water transport and support and are cytologically distinctive cells. They differentiate in the primary body from procambium and in the secondary body from cambium, and because the course of differentiation is generally similar in both cases, it usually will not be necessary to make distinctions between them in the following description.

Both the procambium and the fusiform initials of the cambium appear to be homogeneous in respect to their cellular composition, and do not give evidence of the numerous cell types such as tracheary elements, fibers, sieve tube members, companion cells, or parenchyma that will subsequently be differentiated from them. Therefore, it is reasonable to think that the events leading to the final differentiation of these cells as tracheary elements do not occur much before the first visual evidence of differentiation in this direction. At the level of light microscopic observation the first evidence of differentiation is the enlargement, often both transverse and longitudinal, of the cells. Subsequently the wall becomes increasingly thick by the deposition of additional wall substance against the inner face of the preexisting primary wall (Fig. 14.1). Deposition of this secondary wall may begin before expansion of the cell is complete, and it occurs in highly distinctive patterns, particularly in tracheary elements differentiating from procambium where the secondary wall may consist of a number of discrete bands circling the cell, or one or more spirals traversing the long axis in helical fashion (Figs. 14.2). In other tracheary cells the secondary wall is deposited more extensively and

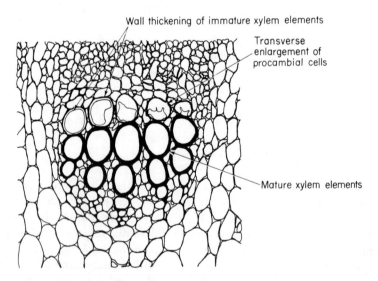

Wall thickening of immature xylem elements

Transverse enlargement of procambial cells

Mature xylem elements

Fig. 14.1   Transverse section of a differentiating vascular bundle in the stem of *Helianthus annuus*. In the center of the bundle the expansion of procambial cells as they differentiate as xylem and begin to lay down secondary wall material can be seen. ×95.

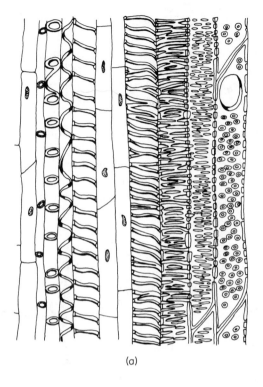

(a)

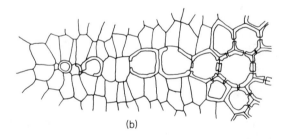

(b)

Fig. 14.2 A sequence of tracheary cells in the stem of *Lobelia* showing the variety of patterns of secondary wall deposition. The cells are named according to the secondary wall pattern and from left to right are : annular (rings of secondary wall material) ; helical (spirals) ; scalariform (ladderlike) ; reticulate (netted) ; pitted. Thin walled cells in the sections are xylem parenchyma. (a) Longitudinal section. (b) Transverse section. (A. J. Eames and L. H. McDaniels, 1947. An Introduction to Plant Anatomy. McGraw-Hill, N. Y.)

forms a network over the inner surface of the primary wall or, as in some cells in the primary body and all those differentiating from cambium, it may form an essentially continuous layer between the primary wall and the protoplast. In these cells the secondary wall is not formed in certain regions where there are dense accumulations of submicroscopic cytoplasmic strands, called *plasmodesmata*, running through the primary wall and establishing continuity between the protoplasts of contiguous cells. These thinner regions become especially conspicuous in tracheary elements with continuous secondary walls and are then called *pits*. The final stages of tracheary differentiation involve chemical specialization of the wall by deposition of lignin within the previously formed cellulose framework. This is followed by lysis and death of the protoplast, and in the case of cells differentiating as vessel members there is also lysis of the end walls to form a continuous multicellular water-conducting tube or vessel (Fig· 14.3). The mature, functioning cells are, therefore, dead, and

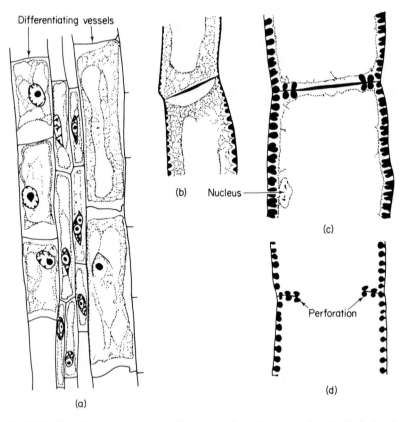

Fig. 14.3 The differentiation of vessel elements in *Apium graveolens* (celery) showing (a) enlargement of the cells, (b) early stage in deposition of the secondary wall, and (c–d) thickening and subsequent dissolution of the end wall to form the perforation, and death and lysis of contents of the cell. (K. Esau, 1936. Hilgardia **10**: 479.)

it is a matter of some interest that death of a cell may be of developmental significance. In animal development morphogenetic cell death contributes to shaping of the extremities of the embryonic limbs, but such cases are rare in plants where cell death seems to be associated more with functional specialization.

The cytological events associated with differentiation of tracheary elements have been studied in more detail recently by high resolution examination made possible by electron microscopy. Such studies have shown that the primary wall, like the primary wall of other cell types, contains an assemblage of randomly-oriented cellulose microfibrils. In contrast, the microfibrils of the secondary wall layers are high oriented, and encircle the cell in helical fashion. This has led to consideration of how such an oriented assemblage may be formed. In an examination of this question in procambial cells of *Avena*, which were differentiating as tracheary elements, Cronshaw and Bouck (1965) found that paralleling the oriented microfibrils in the wall were elongated microtubules in the peripheral layers of the cytoplasm (Fig. 14.4). These occurred only in the regions where a secondary wall was being deposited, and this exclusivity led to the conclusion that the close association between oriented microtubules in the cytoplasm and oriented microfibrils in the wall may indicate a functional significance in wall deposition. At the time when secondary walls were being deposited they also found that endoplasmic reticulum was abundant in the cytoplasm and that the Golgi system was especially abundant and appeared to be actively forming vesicles thought to contribute material to the wall.

Although this study indicated a possible relationship between components of the cytoplasm and the deposition of oriented wall structures, it did not reveal why the secondary wall in some tracheary cells is not deposited as a uniform layer but only in the form of discrete bands or spirals or as a network. Some light has been shed on this question by recent studies, although it is still essentially unresolved. In tracheary cells that differentiated from pith cells after wounding of the stem of *Coleus*, Hepler and Newcomb (1964) found an accumulation of denser cytoplasm in regions where a secondary wall would later be deposited. This cytoplasm was especially rich in Golgi membranes and vesicles, endoplasmic reticulum, and mitochondria; and it was thought that these components were involved in depositing wall material in their vicinity. On the other hand, in sieve tube members and in pollen grains undergoing deposition of additional wall material, thin regions persist that will form pores in which a sheet of endoplasmic reticulum lying in the peripheral cytoplasm becomes closely appressed to the wall. This appears to block deposition of additional wall material in these areas.

It is clear from this brief survey of the differentiation of tracheary

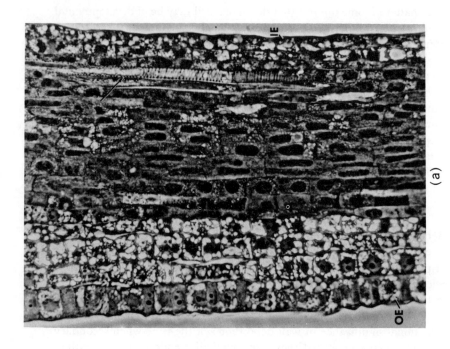

(a)

Fig. 14.4 Tracheary cell differentiation in the coleoptile of *Avena sativa* (oat). (a) Light microscope view of a longitudinal section showing newly differentiated and differentiating tracheary elements with the secondary wall seen in surface view and in section (arrow). (b) Electron micrograph showing part of a differentiating tracheary cell cut in oblique longitudinal section. Secondary wall is deposited inside the primary wall and is seen in section along the sides of the cell and in face view at the top where the plane of the section approaches the cell surface. (c) A glancing section of one of the bands of the secondary wall showing the orientation of the microfibrils in the wall and the similar orientation of microtubules in the adjacent cytoplasm. Key: ER, endoplasmic reticulum; IE, inner epidermis of coleoptile; Mi, microtubule; OE, outer epidermis; Pd, plasmodesma; T, secondary wall; V, vacuole. a ×348, b ×9,990, c ×24,975. (J. Cronshaw and G. B. Bouck, 1965. J. Cell Biol. **24**: 415.)

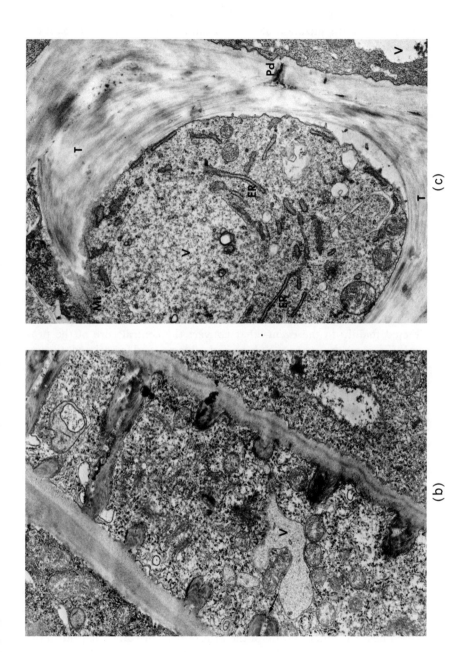

(c)

(b)

223

elements that there are numerous developmental problems to be solved at this level of organization also. For example, the patterned deposition of the secondary wall and the selective lysis of the end wall of vessel members present problems of developmental control that must be answered. However, these questions go beyond the scope of this book, and, however interesting they are, we must leave them to the cell biologists for resolution.

Later stages of differentiation in roots

After consideration of some of the changes that occur in individual cells during the process of differentiation, it is appropriate now to consider once again the higher levels of organization within which these changes occur. As with the initial stages of differentiation, consideration will be given first to the final stages of differentiation in roots where the events are not complicated by the presence of lateral organs.

The late stages of differentiation, involving the maturation of tissues, have been studied in the roots of numerous species. Torrey's study (1955) of differentiation in the root of *Pisum* is a good example, combining structural description with experimental evidence. Attention will be directed mainly to the events that convert the central core of the procambium into the mature xylem and phloem and associated tissues. In the seedling roots of *Pisum* the mature vascular system has a so-called triarch arrangement in which there are three symmetrically arranged ridges of primary xylem in the central core. The establishment of this pattern in the procambial cylinder occurs within 0.2 mm of the promeristem by transverse enlargement of future xylem cells in each of the ridges. The differential enlargement of certain cells occurs first near the center of the root and progresses outward toward the tips of the ridges, or arms as they appear in transverse section (Fig. 14.5). Meanwhile, other procambial cells continue to divide longitudinally and retain a small transverse diameter. In *Pisum* the difference in diameter between cells in the center and at the margin of the central cylinder is not very great, but in some other species the cells that will differentiate as the central xylem vessels attain a very large diameter immediately behind the promeristem.

The final maturation of procambial cells as functional vascular elements occurs first in regions lying between the arms of the future xylem, that is, in the phloem where the first cells to mature are conducting sieve elements located near the outer margin of the procambial core and adjacent to the pericycle (Fig. 14.5). The small group of elements that differentiate in this region constitutes the *protophloem*. Subsequently additional sieve elements differentiate deeper in the procambial core. This constitutes the

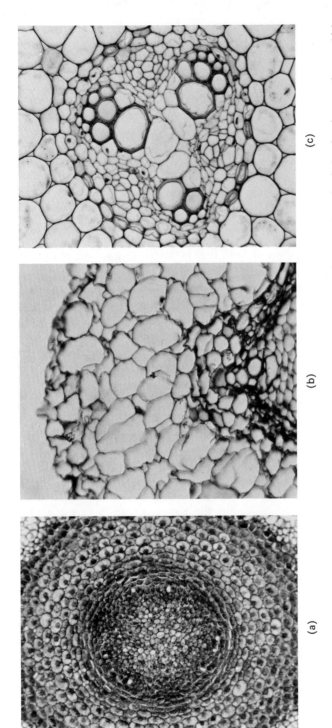

(a)

(b)

(c)

Fig. 14.5   Later stages in the differentiation of vascular tissues in the root. (a) Cross section of the root of *Pisum sativum* (pea) showing the central procambial cylinder in which the first protophloem cells have differentiated at the margin, and in which cells that will differentiate as metaxylem in the center of the cylinder have started to expand. (b) Cross section of part of a pea root showing a later stage of differentiation in which the protoxylem cells have matured and now have lignified secondary walls but the metaxylem cells (at the bottom of the section) are still immature. (c) The central cylinder in the root of *Ranunculus* (buttercup). Protoxylem has differentiated at three points in the cylinder, and protoxylem cells can be identified by their thick lignified walls. Inside each group of protoxylem cells the first metaxylem cells are maturing. These differ from the protoxylem in their larger diameter. The central metaxylem still has thin-walled cells that are immature. The primary phloem lies on radii alternating with the protoxylem. (a) ×42, (b) ×51, (c) ×130.

metaphloem. The terms *protophloem* and *metaphloem* are used to designate a time sequence in differentiation, and they are most valuable when used in this sense without other histological connotations.

The maturation of procambial cells as xylem in the root of *Pisum*, as in other species, occurs some distance proximal to the earliest mature phloem cells, and it is to be found first at the tips of the arms or ridges in the procambial core (Fig. 14.5). This is called the *protoxylem*, and ordinarily consists of elements with a relatively small transverse diameter. Subsequently xylem differentiation proceeds toward the center of the procambial core. The later formed xylem is designated *metaxylem*; and as in the case of the phloem, the prefixes *proto* and *meta* designate a temporal relationship only. The transverse sequence of final xylem differentiation poses an intriguing problem in that it progresses from the periphery to the center whereas the earlier blocking out of future xylem elements by transverse enlargement of procambial cells occurs in the opposite direction. One might expect the first elements that begin the process to be the first to complete it, but in fact they are the last. Undoubtedly the relatively small diameter of the protoxylem elements in comparison with that of the central metaxylem cells is related to the duration of the transverse enlargement phase that precedes the deposition of a secondary wall and the death of the protoplast. In some species, notably among the monocotyledons, the differentiation of xylem does not proceed completely to the center of the root, and a pith of varying dimensions may result from the differentiation of central cells into parenchyma or into non-vascular sclerenchyma. When this occurs, the xylem arms or ridges have the form of separate bundles or strands alternating on separate, equally spaced radii with corresponding bundles of phloem. In large roots of monocotyledons the number of such strands may be 100 or more.

The process of differentiation in the root is a continuous one that proceeds steadily toward the apex without discontinuities (Fig. 14.6). Although the tissues are progressively more highly differentiated as one proceeds from the apex of the root to the base, it is evident that the changes are not related simply to the age of cells. If this were the case, one would expect to find uniform degrees of differentiation at any level of the root, at least within any particular tissue. Clearly this is not the case, because in both xylem and phloem there is a distinct transverse temporal pattern of differentiation. This point will be important later in considerations of the control of differentiation processes.

The outermost cells of the procambial core differentiate in a highly distinctive fashion, and at a level very close to the meristem they can be recognized by their cuboidal shape. The *pericycle*, as this layer of cells is designated, is composed of cells that appear morphologically to be typical parenchyma, but the cells possess distinctive developmental features.

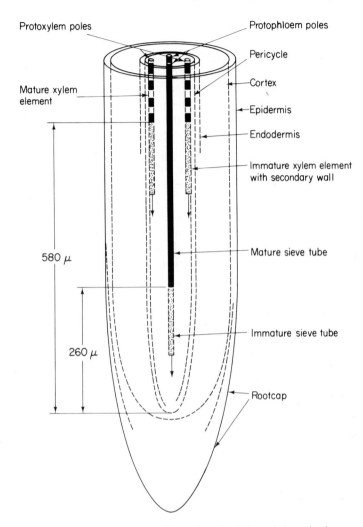

Fig. 14.6 Three dimensional diagram illustrating the differentiation of primary vascular tissues in the root tip of *Nicotiana tabacum* (tobacco). Phloem and xylem both differentiate continuously and acropetally with the phloem in advance of the xylem. (K. Esau, 1941. Hilgardia **13**: 437.)

Whereas other differentiated tissues of the root achieve a relatively stable mature state, the pericycle retains the capacity to initiate lateral roots, to contribute to the origin of the vascular cambium and the cork cambium, and, in some long-lived roots, to give rise to a considerable amount of parenchymatous tissue. It may also be recalled that, in species in which polyploidy occurs in some of the mature cells of the root, the increase in chromosome number does not occur in the pericycle, which remains diploid.

One of the most interesting tissues of the root is the endodermis which ordinarily consists of a single layer of cells lying immediately outside the pericycle. It is developmentally related to the cortex, and the interface between the pericycle and the endodermis marks the boundary between the vascular core and the cortex of the root. Mature endodermal cells are characterized by the presence of a Casparian strip, a narrow band of suberized material running round the cell on its radial and transverse walls, and extending out through the intercellular cementing substance that attaches adjacent endodermal cells. The overall effect of the Casparian strips is thought to make the walls impervious to the passage of water, thus forcing transport through the protoplasts of the endodermal cells, and this layer has figured prominently in considerations of water uptake and transport. The endodermis is relatively late to differentiate and does not appear as a distinctive tissue until after the first xylem has matured. Before any specific morphological features may be detected, endodermal cells acquire distinctive biochemical characteristics which are the result of interaction of substances originating in the vascular system with those having their origin in cortical parenchyma cells. Under conditions of high oxidation in the endodermal cells unsaturated fats are oxidized and deposited in the wall as the Casparian strip. This reaction is subject to experimental manipulation and several environmental factors have been shown to influence the presence or absence of Casparian strips in some species (Van Fleet, 1942).

Later stages of differentiation in shoots

The final differentiation of the vascular system in the shoot consists of the conversion of procambium blocked out behind the apical meristem into mature elements of the xylem and phloem. This process has been studied in a number of species; Esau's investigation of *Linum* (1943) in which the maturation of these tissues has been correlated with the phyllotactic pattern of the shoot is a particularly good example which seems to be typical. It has already been shown in Chapter Thirteen that the procambium of *Linum*, as of most seed plants, forms a system of interconnected strands or leaf traces. Each of these behaves as a unit as far as the timing of its differentiation phenomena is concerned, and it is therefore easier to describe the sequence of events occurring in individual strands than to deal simultaneously with the entire system of the shoot. The overall pattern of final differentiation is illustrated in Fig. 14.7 for tobacco, which is essentially similar to *Linum* in this respect.

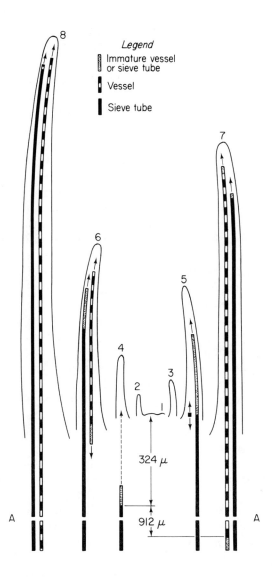

Fig. 14.7   A diagrammatic representation of the differentiation of the first phloem and xylem in the stem and leaves of *Nicotiana tabacum* (tobacco). The leaves are drawn as if they arose in two ranks rather than in a helical pattern. Distances below the shoot apex are shown on the diagram and a segment of the stem has been omitted at A–A. The procambial stage of differentiation is not shown. Mature vascular elements are absent from the leaf traces to the three youngest leaves. (K. Esau, 1938. Hilgardia **11** : 343.)

The first cells to mature in a leaf trace are sieve elements of the phloem (See Fig. 13.6), and the final maturation of phloem progresses acropetally into and through the strand from its basal connection with other leaf traces. This process does not take the form of a uniform maturation of all of the phloem procambium but initially involves only a single file of cells at the outer face of the procambial strand. Subsequently other files of procambial cells located successively deeper in the bundle undergo final differentiation and become mature. Thus, in addition to the acropetal wave of maturation as seen longitudinally, there is a centripetal progression as viewed transversely at any level in the bundle and the primary phloem forms a continuous system throughout its development. The first cells to mature at any level are designated *protophloem*, the later elements *metaphloem*.

The picture that has emerged from the study of xylem maturation in a number of seed plants is quite different and considerably more complex. In *Linum*, as in all seed plants so far investigated, xylem maturation is delayed until after the start of phloem maturation. In *Linum*, the first phloem was found in the traces of leaves just over 0.25 millimeter in length, but the first xylem did not appear in traces of leaves much less than 1.0 millimeter in length. Moreover, the first xylem was found in the leaf itself, near its base, and was not connected by mature elements with fully differentiated elements at lower levels in the stem. Subsequently xylem maturation proceeded both acropetally and basipetally in the trace, ultimately establishing a connection with the mature system. Thus the primary xylem system, although continuous when mature, is discontinuous during its development. In addition, at any level in the trace the first mature xylem elements, the *protoxylem*, were found at the inner face adjacent to the pith and the subsequent maturation of *metaxylem* progressed outward through the procambial trace (see Fig. 13.6). The centrifugal differentiation of xylem, together with the centripetal differentiation of phloem, progressively restricts the remaining procambium. In monocotyledons and some dicotyledons all the procambial cells ultimately differentiate as xylem and phloem, but in many dicotyledons this does not occur and the remaining procambial cells give rise to cambium. The transformation of procambial cells into the cambial meristem which forms the secondary body of the plant will be considered in more detail in Chapter Fifteen.

The pattern of vascular maturation in *Linum*, although complicated to describe, is simple because each leaf is served by only one trace. Where several traces enter each leaf, there are usually differences in the timing of differentiation in each of the traces. Ordinarily differentiation in the median trace is in advance of that in the laterals, and where there are several lateral traces there is usually a median-to-lateral progression of maturation. This considerably complicates the pattern of differentiation seen in transverse sections of the stem because of the great variation in the degree of maturation among the bundles. In monocotyledons, which characteristically have many traces entering each leaf, although the median trace differentiates according to the pattern already described, the lateral traces may show basipetal and discontinuous differentiation, not only of protoxylem and protophloem, but even of procambium (Kumazawa, 1961).

The final stages of differentiation in the shoot of many of the lower vascular plants differ considerably from those found in the seed plants. In *Lycopodium*, which was recently reinvestigated by Freeberg and Wetmore (1967), procambium forms a solid core in the stem and there is no pith. The maturation of procambium as protoxylem and protophloem occurs in the peripheral region and metaxylem and metaphloem are differentiated later in deeper parts of the procambial core. Thus, in the absence of discrete leaf traces, and in the pattern of centripetal maturation of xylem, the stem of *Lycopodium* is remarkably root-like in its development.

In general, investigations of xylem and phloem differentiation have been carried out by traditional histological methods, without regard to seasonal or even diurnal variations. When an attempt was made to consider possible variations of this sort, interesting new information was obtained. By careful round-the-clock sampling in *Coleus*, Jacobs and Morrow (1957) found two isolated centers of protoxylem initiation rather than one. One of these was located, as in other cases, near the base of the leaf, and the other was found further down the strand in the stem. Protoxylem at the lower site was initiated before that at the leaf base, but the two loci became connected so rapidly by differentiation of intervening procambial cells that only closely spaced collections could reveal them as initially separate. It was only in collections made at night that the two centers were observed. In the absence of similar information about other plants one can only wonder whether this situation is peculiar to *Coleus* or may be of more general occurrence.

The evidence presented thus far puts emphasis upon the early discontinuous and basipetal pattern of primary xylem differentiation. There are, however, descriptions of an acropetal progression of xylem differentiation that extends from the mature system below up into the leaf trace and meets the descending basipetal wave of differentiating xylem. Esau observed such acropetal maturation in *Linum*, and it has also been described in *Coleus* by Jacobs and Morrow (1957) and in *Ginkgo* by Gunckel and Wetmore (1946). In other cases where xylem maturation has been studied, no mention of this acropetal wave is made; but because most of these studies have been concerned with only the earliest stages of vascular differentiation in the vicinity of the node of young leaves, an acropetal wave having its origin relatively far down the stem might easily be overlooked. The significance of the acropetal differentiation is not clear. There is certainly a general impression of overall acropetal progression of metaphloem and metaxylem maturation and of cambial initiation where this occurs. This seems to be superimposed upon the protoxylem discontinuities. However, such an impression could be misleading, and despite the numerous studies made of vascular differentiation in the shoot and the considerable amount of factual information now available, it is clear that numerous unanswered questions remain for further study.

In the few cases in which lower vascular plants have been investigated, the view has emerged prominently that acropetal differentiation of both xylem and phloem is frequent and may be extensive in these plants. In the fern *Osmunda cinnamomea*, which has a vascular system consisting of a series of anastomosing bundles surrounding a central pith, a conspicuous acropetal wave of xylem differentiation meets the basipetal wave of protoxylem in the leaf trace. In *Lycopodium*, it was found by Freeberg and Wetmore (1967) that there is extensive acropetal differentiation of both xylem and phloem in the stem and that maturation of xylem precedes that of phloem at any level. However, in *Equisetum* both xylem and phloem are initiated discontinuously at the nodes and connections with older mature vascular tissues are established by basipetal differentiation, with no evidence of acropetal differentiation of either tissue (Golub and Wetmore, 1948).

Experimental studies of vascular differentiation

The relative roles of the shoot apex and leaf primordia in the delimitation of provascular tissue and the differentiation of procambium have been discussed in Chapter Thirteen. The patterns of final maturation of xylem and phloem which have now been described would suggest that in the later stages also there is ample opportunity for the investigation of controlling mechanisms. The differences in pattern in different organs and in different groups of plants may prove useful in such investigations and could provide the basis for comparative experimental studies. However, at the present time it seems that the greatest need is for still more adequate documentation of the patterns of differentiation that actually occur in a variety of plants, and in the absence of such information it is often difficult to interpret the experimental results adequately.

The apparent relationship between the final maturation of vascular tissues and the development of leaves has been the point of departure for a number of experimental studies of the differentiation of xylem and phloem. It has been known for some time that if a vascular bundle in the stem of *Coleus* is severed by a cut, a connection is reestablished between the cut ends of the bundle by the differentiation of intervening pith parenchyma cells into a strand of xylem elements (Fig. 14.8). Jacobs

Fig. 14.8   The differentiation of wound tracheary elements around an incision that severed one of the vascular bundles in the stem of *Coleus*. One end of the severed vascular bundle can be seen at the top right of the figure. The cells in it are elongated and aligned. The regenerated cells are short, they are not highly aligned, and the course of the regenerated bundles through the wound is irregular. ×70. (L. W. Roberts and D. E. Fosket, 1962. Bot. Gaz. **123**: 247.)

(1952) has used this observation as the basis for a series of experiments on the control of xylem differentiation. Removal of the terminal bud and expanding leaves distal to the wound had a marked effect in reducing the amount of xylem differentiated around the wound, whereas removal of leaves and buds from that part of the stem below the wound had little or no effect. The effect of the distal organs could be attributed principally to the expanding leaves which are the major source of auxin in the plant. That the effect of the leaves was actually due to the auxin they produce was demonstrated by the application of indoleacetic acid through the bases of the petioles of leaves that had been removed. In fact, a quantitative relationship between the amount of auxin applied and the extent of xylem differentiation was demonstrated. Furthermore, most of the xylem differentiation proceeded in a basipetal direction, paralleling the predominantly basipetal direction of polar auxin transport in the stem; but a limited amount of acropetal differentiation that occurred could be correlated with a correspondingly small amount of acropetal auxin transport.

In evaluating these experimental observations, Jacobs has commented upon their possible bearing on the isolated centers of xylem differentiation at the bases of developing leaves in intact plants and basipetal differentiation of xylem generally. Such a pattern might be expected if auxin is required for final maturation of xylem and the leaf is the primary source of this substance. On the other hand it is important to remember that the tracheary elements that develop in the experimental plants do so from previously mature parenchyma cells caused to redifferentiate in response to a new stimulus, whereas xylem differentiation in the intact plant is the culmination of normal procambium development. Roberts and Fosket (1962) have pointed out the consequence of this fact in demonstrating that the tracheary cells developing in wounded plants, the wound-vessel members, have a characteristic morphology unlike that of normal vessel members. It is therefore pertinent to cite the evidence that auxin plays a role in normal xylem differentiation from procambium. Jacobs and Morrow (1957) have demonstrated that there is a direct correlation between the rate of xylem maturation from procambial cells and auxin production by the leaf associated with a particular vascular strand. They have also calculated that approximately ten times as much auxin is required to convert a parenchyma cell to a wound-vessel member as is needed to bring about the maturation of a procambial cell into a normal vessel element. This fact, together with the demonstration by Wangermann (1967) that externally applied radioactive indoleacetic acid is transported principally in the vascular bundles of *Coleus*, may explain why pith parenchyma in the intact stem does not differentiate as xylem. It is only after the severing of a vascular strand allows the local accumulation of auxin in excess of normal levels in the region above the wound that the concentration reaches sufficiently high levels to initiate redifferentiation of parenchyma.

Direct evidence for the participation of auxin in the conversion of procambial cells to tracheary elements was obtained by Steeves and Briggs (1960) in experiments on the expanding fronds of the fern *Osmunda cinnamomea*. Excision of pinnae or leaflets removes the source of auxin that normally moves in a polar fashion downward through the rachis and mediates elongation and uncoiling of the leaf axis. In the operated leaves it was also noted that there was a marked reduction in the number of procambial cells converted to mature tracheids. The application of auxin, which restored the expansion and uncoiling of the rachis, also resulted in the complete differentiation of procambial cells as tracheids.

From these experiments it is suggested that hormones of the auxin type are implicated in the final maturation of xylem, but it is not clear to what extent auxin interacts with other factors within a complex system of control. Because of the difficulties of isolating the system in the whole plant, several attempts have been made to examine the phenomena of vascular differentiation in excised plant parts or tissue cultures *in vitro*. It is possible to maintain callus cultures of a number of dicotyledonous species on a medium that will support their growth without the differentiation of vascular tissues. Wetmore and his associates (Wetmore and Sorokin, 1955; Wetmore and Rier, 1963) have experimented extensively with such tissues, particularly of *Syringa*. When vegetative buds were grafted into callus cultures of the same species, nodules and short strands of xylem differentiated, often in a ring-like pattern, in the callus below the grafted shoot (Fig. 14.9a). This suggested the transmission of a stimulus—probably chemical—from the bud into the callus, which had the effect of promoting xylem differentiation. Subsequently it was found that the bud could be dispensed with if in its place agar containing the auxins indoleacetic acid or naphthaleneacetic acid was inserted into the

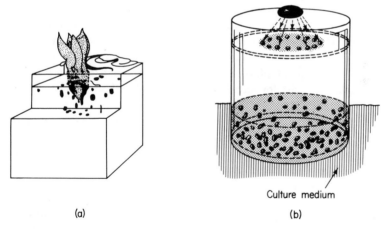

Culture medium

(a)   (b)

Fig. 14.9 Stereodiagrams showing induced vascular tissue differentiation in callus of *Syringa vulgaris* (lilac). (a) Differentiation of xylem in nodules and short strands induced by a grafted bud. Vascular tissue occurs in the wedge-shaped base of the bud and in continuity with this in the adjacent callus tissue. (b) Differentiation of vascular tissue induced by chemical treatment of callus. The culture medium and the agar applied to the top of the callus tissue contain the auxin naphthalene acetic acid and sucrose. Vascular nodules containing both phloem and xylem differentiated in relation to both sources of these substances. (a. R. H. Wetmore and S. Sorokin, 1955. J. Arnold Arbor. 36 : 305; b. R. H. Wetmore and J. P. Rier, 1963. Amer. J. Bot. **50** : 418.)

callus. The distribution of the vascular nodules in the callus could be regulated by the concentration of the applied auxin. When low concentrations were supplied the nodules differentiated close to the auxin source and were progressively further removed with higher concentrations. An interesting further development in these studies was the discovery that simultaneous application of sucrose and auxin to the callus enhanced vascular differentiation (Fig. 14.9b). In these experiments both xylem and phloem were detected in the resulting nodules, often with a cambium between them, and, in fact, the relative proportion of auxin and sucrose applied regulated the relative amounts of xylem and phloem that differentiated. High proportions of sucrose favored phloem development whereas an increase in auxin concentration shifted the balance in favor of xylem differentiation in the treated tissue. As might be expected, in experiments in which a bud was grafted into the callus, if the bud was allowed to develop for a considerable period so that its leaves expended, both xylem and phloem were found to have differentiated in the callus below.

General comment

The suggestion emerging from this work on a control of final vascular maturation dependent upon an auxin-carbohydrate balance is an attractive one when considered in light of the facts of normal differentiation. The discontinuous xylem maturation in the leaf trace could reflect the dependence of this process upon auxin from the leaf, and its delay until acropetal phloem maturation has reached that point may indicate the necessity for an adequate supply of carbodydrate transported by the phloem. The fact that phloem differentiates acropetally and in continuity with mature sieve elements is in agreement with the demonstration in experimental systems of the major importance of carbohydrate for this process. However, auxin is also necessary for phloem differentiation as has been indicated by the tissue culture experiments of Wetmore and Rier (1963) and also by the work of LaMotte and Jacobs (1963) in *Coleus* in which the regeneration of phloem around a wound was shown to be auxin-dependent. Thus the different patterns of xylem and phloem maturation found in the stem are understandable in terms of the distribution of substances known to be present in the plant.

Although the vascular tissues differentiated in tissue cultures include both xylem and phloem, their organization in nodules and sometimes in short strands is very different from that of the system of vascular strands in a stem or even from that of strands regenerated around a wound. Interestingly, Clutter (1960) found that pith plugs of tobacco growing *in vitro*, when supplied with auxin through micropipettes inserted into the tissue, also differentiated nodules of vascular tissue rather than vascular strands. Wetmore and Rier (1963) have noted that the calluses they studied show no polar auxin transport so that auxin movement in them must be by diffusion. Further, Roberts and Fosket (1962) found that xylem regeneration in flowering shoots of *Coleus* and in isolated stem segments of the same species was notably less oriented than in vegetative shoots. This was correlated with reduced polar auxin transport, and the suggestion was made that polar transport may be an important factor in auxin-mediated vascular differentiation in normal shoots.

However, it must be emphasized again that the organization of the vascular system in stem or root does not depend principally upon the processes of final maturation but rather upon the initiation of provascular tissue and of procambium. As yet very little is known about the biochemical factors that control these early phases of differentiation, and one can only speculate as to whether an auxin-sugar balance plays any role in these primary steps. The fact that induced vascular differentiation in callus cultures and in isolated pith plugs is preceded by cell divisions, although not by cell elongation, suggests that the inducing factors influence more than final differentiation. Indeed Fosket (1968) has shown that the differentiation of wound-vessel members in isolated segments of *Coleus* stems does not occur if cell division is blocked by treatment with colchicine. It is evident that investigations directed specifically toward control mechanisms of the early stages of vascular differentiation are required before reliable conclusions can be drawn. Unfortunately, there are few such studies, but experiments on vascular differentiation in roots indicate the direction investigations might follow. In Chapter Twelve an account was given of experiments on the regeneration in culture of *Pisum* root apices following excision of the tips. Torrey (1957) found that the new root formed by a regenerated apex often had a vascular pattern different from that of the original root, particularly with regard to the number of protoxylem poles, thus indicating the relative autonomy of the root apex in controlling the differentiation of its own products. Torrey further noted that the vascular pattern in regenerated roots could be influenced by constituents of the culture medium, particularly its content of indoleacetic acid. When regeneration occurred in a medium containing a high concentration of auxin, the number of xylem poles was increased and the increase was maintained as long as the roots were grown in this medium but reverted slowly to the normal condition following transfer to an auxin-free medium. Measurements showed that the procambial cylinder of roots grown in the high auxin medium was wider than that found in roots in an auxin-free medium. This led Torrey to conclude that auxin, whether occurring endogenously or supplied externally, might be important in controlling the frequency and orientation of cell divisions in the region of the root where the vascular pattern is first established. Thus in a different system auxin is again implicated in vascular differentiation, and in this case it is the initial stages of organization that are regulated. This conclusion is in agreement with results, described in Chapter Thirteen, obtained by McArthur (1967) who found that in isolated, defoliated shoots of *Geum* externally applied auxin caused increased differentiation of provascular tissue, but that for procambium to differentiate either leaves had to be left to develop on the shoot or sugar had to be supplied in the presence of auxin.

In concluding this commentary it is hardly necessary to point out that only the merest outline of the course and causes of differentiation in the plant body has been presented. The very complexity of the processes involved, even in the simplest descriptive terms, is so great as to require dissection into discrete parts for analysis. The description and analysis of the cellular events that apparently are of importance in differentiating tracheary cells of the xylem are likely to be different from those required for the phloem sieve elements, and these will again be different from analyses of differentiation in other cell types. However, the danger of dissecting the phenomenon of differentiation into studies of a number of distinctive cell types is that one may overlook the central question, "How do all of these cell types differentiate in a coordinated way to produce the integrated, functionally efficient organism?"

In searching for answers to this question the possibility of morphogenetic movements of the type described for animal embryos can be eliminated, for these do not occur to any significant extent in plant development. Attention can, therefore, be concentrated on the fact that each cell differentiates in the location where it was formed. There it is subjected to an environment consisting of other cells, and it becomes both the recipient of regulatory stimuli arising in its surrounding milieu and the source of stimuli that may affect the course of differentiation in other cells. While we are now beginning to accumulate information on the respective roles of the terminal meristems and the young leaves in controlling the differentiation of tissues in the plant organs, there is need for additional information on the chemical bases of cellular interactions in development and how such interactions result in the organized development of the plant body. It is to a further examination of this question that Chapter Seventeen is directed.

REFERENCES

Clutter, M. E. 1960. Hormonal induction of vascular tissue in tobacco pith *in vitro*. Science **132**:548–549.

Cronshaw, J., and G.B. Bouck. 1965. The fine structure of differentiating xylem elements. J. Cell Biology **24**:415–431.

Esau, K. 1943. Vascular differentiation in the vegetative shoot of *Linum*. II. The first phloem and xylem. Am. J. Botany **30**:248–255.

Fosket, D. E. 1968. Cell division and the differentiation of wound-vessel members in cultured stem segments of *Coleus*. Proc. Nat. Acad. Sci. U. S. **59**:1089–1096.

Freeberg, J. A., and R. H. Wetmore, 1967. The Lycopsida—A study in development. Phytomorphology **17**:78–91.

Golub, S. J., and Wetmore, R. H. 1948. Studies of development in the vegetative shoot of *Equisetum arvense* L. II. The mature shoot. Am. J. Botany **35**:767–781.

Gunckel, J. E., and R. H. Wetmore. 1946. Studies of development in long shoots and short shoots of *Ginkgo biloba* L. II. Phyllotaxis and the organization of the primary vascular system; primary phloem and primary xylem. Am. J. Botany **33**:532–543.

Hepler, P. K., and E. H. Newcomb. 1964. Microtubules and fibrils in the cytoplasm of *Coleus* cells undergoing secondary wall deposition. J. Cell Biology **20**:529–533.

Jacobs, W. P. 1952. The role of auxin in differentiation of xylem around a wound. Am. J. Botany **39**:301–309.

Jacobs, W. P., and I. B. Morrow. 1957. A quantitative study of xylem development in the vegetative shoot apex of *Coleus*. Am. J. Botany **44**:823–842.

Kumazawa, M. 1961. Studies on the vascular course in maize plant. Phytomorphology **11**:128–139.

LaMotte, C. E., and W.P. Jacobs. 1963. A role of auxin in phloem regeneration in *Coleus* internodes. Devel. Biol **8**:80–98.

McArthur, I. C. S. 1967. Experimental vascular differentiation in *Geum chiloense* Balbis. Thesis, University of Saskatchewan.

Roberts, L. W., and D. E. Fosket. 1962. Further experiments on wound-vessel formation in stem wounds of *Coleus*. Botan. Gaz. **123**:247–254.

Steeves, T. A., and W. R. Briggs. 1960. Morphogenetic studies on *Osmunda cinnamomea* L. The auxin relationships of expanding fronds. J. Exper. Botany **11**:45–67.

Torrey, J. G. 1955. On the determination of vascular patterns during tissue differentiation in excised pea roots. Am. J. Botany **42**:183–198.

——— 1957. Auxin control of vascular pattern formation in regenerating pea root meristems grown *in vitro*. Am. J. Botany **44**:859–870.

Van Fleet, D. S. 1942. The significance of oxidation in the endodermis. Am. J. Botany **29**:747–755.

Wangermann, E. 1967. The effect of the leaf on differentiation of primary xylem in the internode of *Coleus*. New Phytol. **66**:747–754.

Wetmore, R. H., and J. P. Rier. 1963. Experimental induction of vascular tissues in callus of angiosperms. Am. J. Botany **50**:418–430.

Wetmore, R. H., and S. Sorokin. 1955. On the differentiation of xylem. J. Arnold Arboretum **36**:305–317.

FIFTEEN

Secondary Growth—

The Vascular Cambium

In considering the apical or primary meristems of the plant body, one of the most perplexing problems is the permanently meristematic condition of these regions, which are somehow spared from the processes of maturation occurring in their derivatives. One might be tempted to relate this property to their terminal position, their three-dimensional mass, or their organization which is distinct from that of the mature structures that they produce. However, the lateral meristems, which share the capacity for continued growth but are strikingly different in every other respect, prevent an easy aquiescence to this temptation. The *vascular cambium* and the *cork cambium* or *phellogen* are lateral in position, have the form of cylindrical sheets encircling the plant axis, and are organized in close conformity with the tissues to which they give rise. Furthermore they initiate only specific tissues rather than whole organs as in the case of the terminal meristems. In fact, the marked dissimilarity between the terminal and lateral meristems raises the very serious question of the desirability of using a single term, *meristem*, to designate both. Although the term has been defined in such a way as to include both categories, its use may result in obscuring the marked differences between them. Furthermore it must be borne in mind that whereas every vascular plant body must have terminal meristems in order to exist at all, the lateral meristems have a supplemental role and are by no means universal (Barghoorn, 1964).

The initiation of cambial activity

Nothing emphasizes the dissimilarity between primary and secondary meristems more effectively than a consideration of the origin of the vascular cambium. Whereas the shoot and root apical meristems are initiated among the cells of the embryo early in the development of the plant, the cambium has its origin from a partially differentiated vascular tissue, the procambium. Although the major significance of the procambium is that it is a stage in the differentiation of the primary vascular system, certain cells may become stabilized at this level of maturation, retaining the capacity to proliferate indefinitely, and thus give rise to a vascular cambium. The details of cambial origin are best presented as they occur at one level in a stem or root at some distance proximal to the growing tip; but it must be kept constantly in mind that such a picture is misleading because it tends to obscure the continuous acropetal extension of cambial activity. The events to be described, in fact, occur at the distal extremity of the already existing cambium, and in continuity with it.

In the stem of a typical dicotyledon, the development of procambium in a bundle is accompanied by numerous longitudinal divisions. As these proceed, they ordinarily tend to become oriented parallel to the surface of the stem, that is, periclinally, with the result that the procambial cells tend to occur in radial rows. The final maturation of xylem elements begins, at any level, at the inner face of the bundle adjacent to the pith, and phloem maturation begins opposite to it at the outer face. Subsequently the final maturation of these two tissues proceeds toward the center of the bundle gradually restricting the procambium to a band of dividing cells. Ultimately a stage is reached, by no means easily recognizable, at which it may be deduced that a common meristem is producing elements that will ultimately differentiate as both xylem and phloem (Fig. 15.1). At this point it is stated that secondary growth has begun or that cambium has been initiated. Clearly, however, cambial activity represents a continuation of processes begun as part of primary development; there is no sharp demarcation between primary and secondary growth.

Cambial activity is restricted to the bundles in which it is initiated in cases in which there is a very limited amount of secondary growth, but more commonly there is a further development which results in a continuous ring, or more properly a cylinder, of cambium. After the initiation of fascicular cambium—that in the bundles—at any level, partially differentiated parenchyma between the bundles begins to divide periclinally and the interfascicular cambium is initiated (Fig. 15.1). Where the bundles are widely spaced, this activity can be observed to begin

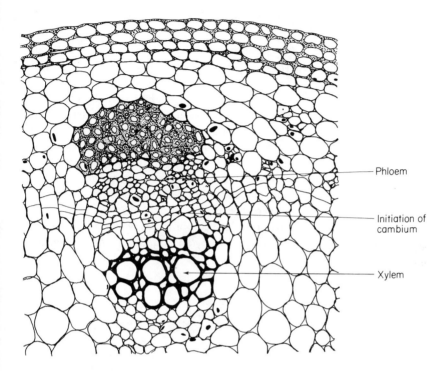

Fig. 15.1   Initiation of vascular cambium in the stem of *Helianthus annuus* (sunflower). Cambial activity that starts in the vascular bundles is beginning to extend into the interfascicular regions as a series of tangential divisions in previously differentiated parenchyma cells. ×215.

adjacent to that in the bundles and to advance across the interfascicular region, suggesting that the stimulus for the renewal of meristematic activity in the interfascicular regions has its origin in the dividing cells of the bundle.

This sequence of events in cambial initiation in stems is generally found in the gymnosperms as well as in the dicotyledons, but in the relatively few monocotyledons that have secondary growth the origin is very different. Here the cambium arises outside the primary vascular tissues and produces both xylem and phloem, in bundles, to the interior. In roots also, the differences from stems in organization and in the pattern of differentiation of primary vascular tissues impose corresponding differences in the manner of cambial initiation. Both primary xylem and phloem differentiate centripetally, that is, toward the center of the root, and along alternating radii rather than in the same bundles. The initiation of cambium occurs in the procambium of the phloem strands adjacent to the xylem, and the cambial ring is completed by the pericycle opposite the protoxylem poles.

The organization of the vascular cambium

In view of what has been said, it is evident that the methods of study of the vascular cambium, both analytical and experimental, must be quite different from those applied to terminal meristems, and the difficulties are even more acute than in the case of the root apex. The cambium may be observed effectively only in sections, which makes the study of living material difficult; and because of the extended, sheet-like form of the cambium and its relationship to its derivative tissues, the sections must be oriented very precisely in order to be meaningful (Fig. 15.2). The general organization of the cambium, as well as the techniques for its investigation, may be illustrated by an examination of this meristem

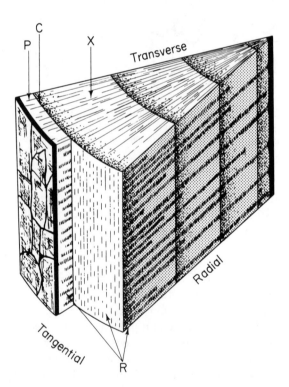

Fig. 15.2   Wedge from the stem of *Pinus* sp. (pine) illustrating the position of the vascular cambium in relation to its derivative tissues, and the orientation of sections used to examine the cambial meristem. Key: C, cambium; P, secondary phloem; R, vascular ray; X, secondary xylem. (Adapted from E. Strasburger. 1921. A Textbook of Botany, Macmillan, London.)

in the stem of *Pinus strobus* (white pine), which was the object of a series of investigations by I. W. Bailey (1954) now regarded as classics in the field.

The cambium of Pinus strobus

The cambium may be observed in face view by cutting sections that are tangential to the surface of the stem after removing the outer tissues overlying it (Fig. 15.2). Because the cambium is a thin layer sandwiched between the secondary xylem and the secondary phloem to which it gives rise, a section of the meristem ordinarily includes some areas of the derivative tissues on either side of the cambium itself. This is due in part to obliquity in the sections and in part to the curvature of the organ under investigation. Bailey has provided good evidence that if sections are cut in this way from living blocks of tissue, the cambium may be observed in a relatively normal condition. Such tissue slices were kept alive, as evidenced by cytoplasmic streaming, for periods of more than two months in ordinary tap water, but no cell division was noted. For the study of cellular detail tissue blocks may be fixed and sections cut and stained by standard histological methods.

When a tangential section of cambium is examined, one is immediately struck by the difference in organization between this tissue and the primary meristems (Fig. 15.3). The tissue is made up of two fundamentally different kinds of cells. The *fusiform initials* are elongate, pointed cells, up to 4 millimeters in length and 0.04 millimeter in width. The visible cell walls are thick and are beaded in appearance because of the occurrence of numerous primary pit fields. The cytoplasm is highly vacuolated and, if living sections are examined, protoplasmic streaming can often be observed. In contrast to the fusiform initials, which have no conspicuous vertical or horizontal pattern of alignment, the *ray initials* are arranged in short, uniseriate vertical rows. These cells are only 0.02 millimeter in height and their width is approximately the same. Comparison with the mature xylem and phloem, cut in the same plane, often visible in the same preparation, reveals a close similarity in histology between the mature and the meristematic tissues. The fusiform initials clearly correspond to the tracheids in the xylem and to the sieve cells of the phloem, whereas the groups of ray initials are related to the horizontal rays that extend through both xylem and phloem.

The details of the spatial relationship between the cambium and its derivative tissues may be examined more fully in transverse sections (Fig. 15.4). Clearly defined radial rows of cells extend through both xylem and phloem, and these two tissues are separated by a zone several cells in width in which, although the rows are distinct, the cells are immature and are apparently dividing. As in the tangential sections, two kinds of cells can be identified but here their aspect is much different.

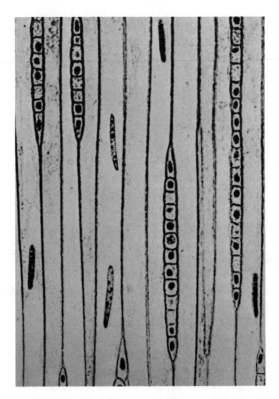

Fig. 15.3   Cambium of *Pinus strobus* (white pine) viewed in tangential section. ×200.
(I. W. Bailey, 1920. Amer. J. Bot. **7**: 417.)

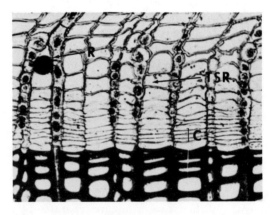

Fig. 15.4   The cambium and its derivative tissues in *Pinus strobus* (white pine) seen in
transverse section. Key: C, cambial zone; R, ray; TSR, tracheid-sieve cell row. ×200. (I. W.
Bailey, 1920. Amer. J. Bot. **7**: 417.)

In the tracheid-sieve cell rows, which must contain the fusiform initials, the narrowest cells have a radial diameter of only 0.006 millimeter and the corresponding cells in the rays have a radial dimension of 0.03 millimeter. It is clear from the orientation of cell walls that cell divisions must occur primarily in a longitudinal, periclinal plane, that is, in such a way as to increase the number of cells in the rows. This division process is not difficult to visualize in the rays; but in the case of the fusiform initials, whose length is more than 600 times the radial width, such a division plane raises some interesting questions. A periclinal longitudinal division in a fusiform initial partitions the cell at right angles to its narrowest dimension and forms a wall having the maximum possible surface area (Fig. 15.5b).

In the foregoing discussion descriptions were given of what were called fusiform and ray initials. In transverse sections taken during periods of active growth, however, no single cell may be picked out in each row as an obvious initial. Rather in each row there are several cells,

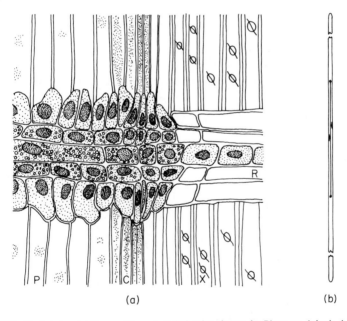

(a)   (b)

Fig. 15.5 Radial view of the cambium and derivative tissues in *Pinus* sp. (pine). (a) The cambium and derivative tissues in radial section. The cambium has elongated fusiform initials and short ray initials. (b) A single fusiform initial undergoing division in the periclinal plane. The nucleus has divided and two daughter nuclei can be seen separated by the new wall which has not yet grown to the tips of the cell. Key: C, cambial zone; P, secondary phloem; R, ray; X, secondary xylem. (a) ×240. (a. Adapted from E. Strasburger, 1921. A Textbook of Botany, Macmillan, London. b. I. W. Bailey, 1920. Amer. J. Bot. **7**: 417.)

any one of which on the basis of position and structure could be regarded as an initial. Some observers have maintained that in each row there must be a permanently meristematic initial cell and that the similar cells on either side of it, although still dividing, are xylem or phloem mother cells. Others have argued that there is merely a zone of dividing cells and that no individuals have any permanent status as initials. A discussion of the merits of these two points of view will be deferred until some aspects of cambial growth have been considered. This problem must be kept in mind, however, in the examination of cambial histology in that what have been designated fusiform and ray initials are not necessarily permanent initials.

Longitudinal sections cut along a radius of the stem complete the picture of the shape of cambial initials (Fig. 15.5). In such radial sections, the long initials and their immediate derivatives are seen to have blunt ends, which facilitates an understanding of the periclinal division process. Furthermore the marked disparity in radial width of fusiform and ray initials is very evident in such sections.

Cambium in other species

The cambium has been investigated histologically in a relatively large number of species most of which, however, are trees. Other conifers differ little from white pine in the structure of the cambium but there is considerable variation in the length of fusiform initials, from less than one millimeter to nearly nine millimeters. In woody dicotyledons, the cambium usually contains multiseriate groups of ray initials of various heights and widths in keeping with the occurrence of multiseriate rays in the secondary xylem and phloem. There is also a much greater range in the length of fusiform initials, and in the vast majority of species these are much shorter than in the conifers (Fig. 15.6a). In some primitive, vesselless dicotyledons, however, the length is as great as it is in the conifers. Bailey has shown that there has been a phylogenetic reduction in the length of fusiform initials in dicotyledons that is correlated with a reduction in the length of vessel elements. In some very advanced species, as judged by other criteria, the very short fusiform initials occur in horizontal tiers in what is designated as *storied* or *stratified cambium* (Fig. 15.6b).

Reference has already been made to the distinctive origin of the cambium in the relatively few monocotyledons that have secondary growth. The structure is also very different from that found in conifers and dicotyledons. The cambial initials are all of one type and vary greatly in shape even in the same plant. They may be either fusiform or rectangular, and in some cases the same cell may have one tapering end and

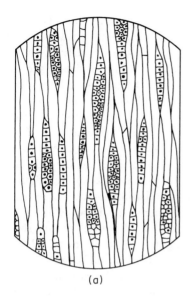 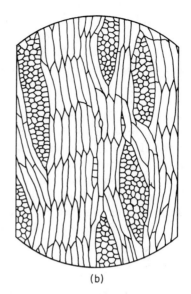

Fig. 15.6   Tangential view of the cambium of *Juglans cinerea* (walnut) (a) and *Robinia pseudoacacia* (locust) (b). There are multiseriate and uniseriate rays in both species, and in *Robinia* the fusiform initials are in a storied arrangement. ×100. (Redrawn from A. J. Eames and L. H. MacDaniels, 1947. An Introduction to Plant Anatomy, McGraw-Hill, N. Y.)

one blunt end. Because of its distinctive structure and activity and its developmental relationship with the meristematic activity which increases the thickness of the primary body, there is some doubt as to whether this meristem is truly homologous with the cambium of conifers and dicotyledons.

The dynamic state of the cambium

The picture of the cambium presented thus far is a rather static one, although the extensive cell-producing activity has been emphasized. The continued formation of secondary xylem, however, imposes upon the cambium the necessity of covering an ever-increasing surface. A picture of the cambium as a constantly changing dynamic population of meristematic cells has emerged from several intensive investigations.

Fusiform initials in conifers

The mechanism by which increase in girth is achieved in the cambium of conifers has been studied revealingly in *Pinus strobus* by Bailey who has shown that a surprising number of different processes are involved. During a period of approximately 60 years from its initiation at any level in a tree of this species, the cambium is characterized by an increase in the size of its fusiform initials, particularly in length and in tangential width. Increase in length in fusiform initials is accomplished by tip growth, and the pointed ends of the cells intrude between other initials. This process, therefore, leads to an increase in the number of fusiform initials intersected in any transverse section and, together with increase in width, contributes to the enlarged circumference of the cambial layer. However, Bailey has shown that these processes cannot explain the total increase in circumference and, because they do not continue beyond approximately 60 years, they cannot explain subsequent increase in girth. Because the number of cambial initials in cross sections does continue to increase, it is evident that there must be anticlinal divisions in the fusiform initials. Tangential sections, however, do not show the expected evidence of such divisions in that the fusiform initials are not arranged in pairs or horizontal groups, which would suggest a common origin. The surprising resolution of this difficulty is found in the observation that anticlinal divisions in the fusiform initials are nearly transverse in orientation (Fig. 15.7a). Bailey has termed such divisions *pseudotransverse*. The result of such a division is the production of two initials, each approximately half the length of the original. Following division, intrusive growth occurs at the tips of each of the initials, both of which finally achieve the length of the original. Thus the effective number of cambial cells in cross section is increased (Fig. 15.8a, b).

Bannan (1956) has provided information on the frequency of anticlinal divisions in *Thuja occidentalis* and several other conifers. Because periclinal divisions that lead to the formation of xylem and phloem derivatives continue during these developmental changes, and because the derivatives elongate only slightly during differentiation, a record is preserved in the secondary xylem in which the time sequence can be determined. A considerable interval occurs between successive anticlinal divisions in the same row, but this is highly variable. In *Thuja* Bannan found that in mature trees the average interval in different trees ranged from 1 to 8 years and that the overall average for all trees studied was 3.7 years. He has found that in the first few years after initiation at any level in the stem, the cambium is characterized by a high frequency of pseudotransverse divisions, and that this frequency declines in successive years until a relatively stable value is reached. This is correlated with the previously noted trend in the increase in length of fusiform initials during the early years of cambial activity (Bannan, 1960). That the inverse correlation between rate of anticlinal division and length of fusiform initials is a real one is further demonstrated by Bannan's (1957a) studies on fluted trunks in which he has shown that in the depression of

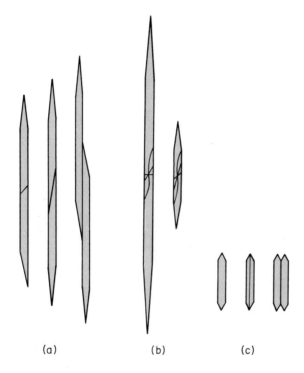

(a)   (b)   (c)

Fig. 15.7 Anticlinal divisions of cambial fusiform initials. (a) Pseudotransverse division and subsequent elongation of the two daughter cells in a conifer. (b) A series in dicotyledons showing the increasing vertical orientation of the anticlinal division associated with decreasing length of the fusiform initial. (c) Vertical anticlinal division that results in a storied cambium. (I. W. Bailey, 1923. Amer. J. Bot. **10**: 499.)

such trunks the rate of anticlinal division is higher than elsewhere and the length of fusiform initials is less. Increase in girth, however, is not directly correlated with the rate of anticlinal division because of the fact that fusiform initials drop out of the cambium by becoming differentiated and thus ceasing to divide. The evidence for such cessation of activity is the termination of a radial row in both xylem and phloem (Fig. 15.8c). Thus the increase in circumference is governed by a balance between the rate of anticlinal division and the rate of drop-out of initials; and in the depressions of fluted stems there may be a net loss of initials even though the rate of anticlinal division is high. The drop-out of an initial can occur at any time in its life history but it occurs most often in the period shortly after an anticlinal division.

In any consideration of the possible regulation of cell division in the cambium, it is necessary to distinguish between periclinal divisions that

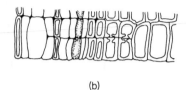

(a)

(b)

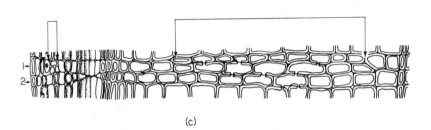

(c)

Fig. 15.8   Addition and loss of fusiform initials in the cambium of *Thuja occidentalis* (white cedar). (a) As the result of a pseudotransverse anticlinal division a fusiform initial of the cambium has doubled. The cambial layer is shown stippled. The secondary phloem (left) and secondary xylem (right) occur as a single row of cells that were produced by the activity of a single initial. The divided fusiform initials have not yet produced derivative cells. (b) A developmental stage in the cambium later than that shown in *a*. The two cambial initials have produced a doubled row of phloem on the left and xylem on the right which is enclosed in the single row produced before the initial doubled. (c) In the row of cells marked "1" the fusiform initial doubled producing a double row of phloem (left) and xylem (right). The doubled parts of the phloem and xylem are indicated by the connected arrows. One of the initials then ceased to function and the xylem and phloem rows became single again. In the cell row marked "2" the fusiform initial has just ceased to function. It is not present in the cambial zone which is stippled. Its derivative xylem and phloem row of cells will now terminate. (M. W. Bannan, 1955. Canadian J. Bot. **33**: 113.)

are overwhelmingly predominant in frequency and anticlinal divisions that, although much less frequent, are essential in maintaining the integrity of the meristem. Bannan has obtained evidence indicating that although anticlinal division can occur throughout the annual growth period in conifers, there is a tendency for it to occur most often toward the end of the period when periclinal divisions are greatly reduced in frequency. This observation might be extremely significant in any experimental attempt to separate the controls of the two patterns of division.

Fusiform initials in dicotyledons

In the woody dicotyledons, Bailey has found interesting differences in the mechanism of girth increase in the cambium which are related to the shortening of fusiform initials. In the primitive vesselless species the process is apparently similar to that in conifers. In a series of species having progressively shorter fusiform initials, he has shown that the anticlinal divisions become increasingly oblique, approaching progressively nearer to the vertical orientation (Fig. 15.7b). As the divisions become more nearly vertical, there is a decreasing amount of instrusive elongation following the divisions. The ultimate stage in this series is one in which the shortest fusiform initials are found, and in these the anticlinal divisions are completely vertical and are followed by no elongation (Fig. 15.7c). This process leads to the formation of horizontal tiers, the storied cambium previously mentioned.

Reference has been made to an ontogenetic trend in the cambium of conifers that leads to an increase in the length of fusiform initials in the early years of cambial activity. In the dicotyledons, if the series of species with decreasing length of fusiform initials is examined, it is found that this ontogenetic trend is correspondingly suppressed and is essentially absent in species having storied cambium. The suppression of this trend is, of course, an important factor in bringing about the overall reduction in length of the fusiform initials. Because this trend is apparently explained by a reduction in the frequency of anticlinal divisions, with a corresponding enhancement of the elongation of the fusiform initials, it is not surprising that its disappearance is correlated with a shift in anticlinal divisions from the pseudotransverse to the vertical orientation and a loss of the ability to elongate following such divisions. It is significant to note, however, that these changes cannot be explained simply as a result of a loss of the ability of fusiform cells to elongate intrusively because, even in those cases in which there is no elongation at all after anticlinal division, derivatives of the cambium that develop as fibers undergo extensive elongation during their differentiation.

Changes in the rays

The foregoing discussion has emphasized that the fusiform initials of the cambium are in a state of constant change. As the secondary body enlarges and the cambium expands, it is evident that the number of rays and consequently the number of groups of ray initials also increase. In fact Bailey has noted the increase in the number of ray initials as a factor contributing to increase of girth in the cambium in both conifers and dicotyledons. In a study of ray ontogeny in conifers and dicotyledons, Barghoorn (1940a, b; 1941) has investigated the methods by which new rays arise and the subsequent developmental changes that they undergo as a result of changes in the cambium (Fig. 15.9). In the conifers

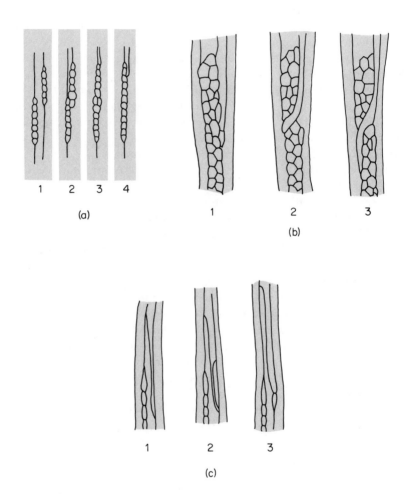

Fig. 15.9 Ontogenetic changes in rays of conifers and dicotyledons revealed by study of the secondary xylem in which the changes are preserved. (a) Fusion of two rays in *Taxus baccata* that results from dropping out of a fusiform initial. (b) Division of a ray in *Trochodendron aralioides* by intrusive elongation of a fusiform initial. (c) Formation of a new ray from the tip of a fusiform initial in *Viburnum odoratissimus*. (b) ×75, (c) ×35. (a. E. S. Barghoorn, 1940. Bull. Torrey Bot. Club **67**: 303; b, c. E. S. Barghoorn, 1940. Amer. J. Bot. **27**: 918.)

a new ray may arise from a fusiform initial by the cutting off of a cell from the tip which then begins to function as a ray initial, by the cutting out of a cell from the side of a fusiform initial by means of an anticlinal division in which the new wall curves to intersect the side wall, or by the septation of an entire fusiform initial to form a vertical series of ray initials. The uniseriate rays of conifers do not increase in width but they increase in height as a result of horizontal anticlinal divisions in ray initials or by the fusion of two rays as the result of the dropping out of a fusiform initial. The number of rays may also be increased by the division of existing rays as a result of penetration by the elongating tips of fusiform initials.

In dicotyledons new rays have their origin from fusiform initials as in the conifers, but there are two differences to be noted. Ray initials ordinarily are not cut out of the side of fusiform initials, and occasionally a multiseriate ray may arise directly by septations involving several fusiform initials. The developmental changes that occur in rays of dicotyledons are numerous and complex and only a few will be noted here. Rays increase in both height and width by division of fusiform initials, and uniseriate rays are often converted into multiseriate rays. Fusion of rays leads to an increase in size, and ray dissection leads to a decrease in size and an increase in the number of rays. Dissection may be accomplished by intrusion of elongating fusiform initials and by the conversion of ray initials to fusiform initials. In view of the highly dynamic state of rays, Barghoorn has pointed out that it is essential to understand their ontogenetic changes before using them in the systematic description of secondary xylem or attempting to establish phylogenetic trends in ray structure.

The question of cambial initials

Up to this point the cambium has been treated as if it consisted of a single layer of initial cells although such a layer is not ordinarily distinguishable from its immediate derivatives. It would seem to be desirable now to examine the evidence favoring this interpretation over the alternative view that the cambium is a multilayered meristem in which there are no cells having any permanent value as initials. The information presented dealing with developmental changes in the cambium provides strong support for the concept of initial cells. It is difficult, if not impossible, to interpret the simultaneous changes occurring in individual radial rows on the two sides of the cambium in terms other than that these changes have occurred in single initial cells that are relatively permanent. Doubling of a row on both sides of the cambium, the loss of a row on both sides, and the various changes in rays occurring in both xylem and phloem are examples of the sort of changes that must be dependent on the existence of permanent initials (Fig. 15.8, 15.9). Occasionally temporary changes occur on one side only. For example, a xylem row may become double over the extent of several cells and then become single again. Such a phenomenon could result from an anticlinal division in a derivative of the fusiform initial in that particular row, a xylem mother cell, which still had considerable division potential remaining and strongly supports the idea that the more common permanent changes are those that occur in the true initials. The functional distinctiveness of the single layer of initials is emphasized by Bannan's (1957b) observations on 12 species of conifers in which 98 percent of the pseudotransverse divisions noted were of the permanent type.

In view of this convincing but indirect evidence for the existence of functionally distinctive initial cells, it is appropriate to examine the cambial region in detail to determine whether any direct evidence may be obtained. Such examination of transverse sections reveals immediately that the periclinal, tissue-generating divisions are not restricted to a single layer of cells but rather are distributed through a multicellular zone. Wilson (1964, 1966) has described this zone in *Pinus strobus* as consisting of 6 to 15 cells in each radial row which are actively dividing periclinally, and Bannan (1955) has found even wider zones in rapidly growing trees of *Thuja*. It is generally agreed that there is no single cell in each row of elongate cambial cells that stands out structurally as the initial of that row. Newman (1956) has pointed out, however, that in *Pinus radiata* the ray initials can be distinguished on the basis of size from maturing ray cells on either side, and this may be true in many species. This same worker has argued that close scrutiny of cell wall structure in the cambial zone permits the identification of the fusiform initial, at least in some rows (Fig. 15.10). His analysis depends upon the fact that each time a cell divides a complete wall layer is deposited around the protoplast. Thus the relative age of walls can be determined to a considerable extent by their thickness, and packets of cells within a common wall can be observed. In *P. radiata* the fusiform initials alternately produce groups of derivatives on xylem and phloem sides and one may speak of xylem-forming and phloem-forming phases of activity. By locating the general area of the initial from the position of the ray initials and looking for a particularly thick tangential wall, Newman felt that he could identify the initial cell as the one bounded by that wall and could tell whether it was in a xylem or a phloem forming phase according to the side on which the wall was located. If it was on the phloem side, he concluded that the initial was in the process of producing xylem derivatives, and if it was on the xylem side the initial was held to be in a phloem-producing phase. From Newman's study it was also apparent that the individual radial rows were not coordinated in their activity so that of

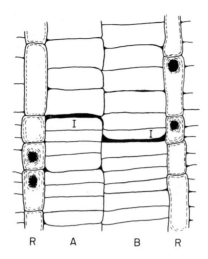

Fig. 15.10 Diagram illustrating Newman's interpretation of the cambial activity in *Pinus radiata* (pine). The approximate location of the initials in the cambium is first identified by the position of the smallest cells in the rays which are assumed to be the ray initials. The fusiform initials will then be approximately next to the ray initials. The fusiform initials can be recognized by the wide band of wall material on their inner or outer tangential face. When the band is on the outer face of the fusiform initial, as in cell row A the initial is in its xylem-producing phase (xylem is at the bottom of the diagram and phloem at the top), and when the band is on the inner face of the fusiform initial it is in its phloem-producing phase as is the case in cell row B. Key: I, fusiform initial; R, ray. (Diagram based on data from I. V. Newman, 1956. Phytomorphology **6** : 1.)

two adjacent rows, at any particular moment one might be forming xylem and the other phloem.

Newman's observations emphasize the important, but sometimes overlooked, fact that the cambial initials do not function like the apical cell of a vascular cryptogam by cutting off derivatives in a sequential pattern. Following periclinal division in an initial, either one of the sister cells may retain the initial function while the other becomes a tissue mother cell. Because adjacent radial rows are not necessarily coordinated in this respect, the concept of a single layer of initials is somewhat misleading if it gives a three-dimensional picture of a smooth cylinder. In fact, except during dormant periods, the layer must be rather irregular because the initials of adjacent rows may be considerably offset.

In plants with marked periodicity of cambial activity, during the dormant period the cambial zone is reduced in width and in some cases the presence of an initial in each radial row becomes more obvious. Esau (1948) has described the cambial zone in *Vitis* as being reduced to a single layer of undifferentiated cells, presumably the initials, during the period of winter dormancy. A more typical situation is that described by Bannan (1955) in *Thuja occidentalis*. In this species, the cambial zone in dormant trees is usually three cells wide. The outermost of these cells, adjacent to recognizable phloem, is regarded as the initial and is distinguished from the inner xylem mother cells which are slightly longer. When growth is renewed in the spring, cell division most commonly begins in the xylem mother cell next to the mature xylem, and somewhat later division occurs in the initial cell. For a time, cell division proceeds more rapidly than cell differentiation so that a broad zone of dividing cells is established. Within this broad zone, however, there is a nonuniform distribution of mitotic activity. The peak of mitotic activity is not found in what Bannan interprets to be the initial cells, but rather occurs near the center of the group of xylem mother cells. Cell division frequency in the initial cells and phloem mother cells is approximately one-half the maximum value.

Terminology

Now that consideration has been given to the organization and functioning of the lateral vascular meristem, it may be appropriate to comment upon the inconsistencies that are apparent in the terminology applied to this region. In some cases the term *cambium* is used specifically to designate the single layer of initial cells even though these are extremely difficult if not impossible to identify. The entire band of dividing cells is then designated the *cambial zone* or *cambial region*. An alternative approach has been to use the term *cambium* to describe the whole dividing region and to speak of *cambial initials* or *fusiform* and *ray initials* when these are intended. The latter terminology, which has been more widely followed, seems to be preferable for several reasons. In the first place, the restriction of the term *cambium* to the layer of initials probably would have little influence upon nonspecialists who ordinarily use it in reference to the whole lateral meristem. Furthermore the broader usage of the term serves to make it more nearly comparable to the terms *apical meristem* or *promeristem*, which are considered to include not just the apical initials but their recent derivatives as well.

General comment

One of the most significant aspects of the study of the vascular cambium is the bearing it has upon our understanding of the mechanism of continued meristematic activity. If one considers origin, position, histological organization, and cytological characteristics, there is essentially nothing common between the vascular cambium and either the shoot apex or the root apex. So great are the differences in fact that it is tempting to speculate that the long retention of meristematic potentialities may have a different physiological basis in different parts of the plant body; but, so long as the actual mechanism remains unknown in all cases, such speculation has little value. The one exception to the general dissimilarity between the primary meristems and the cambium could be the occurrence of permanent initial cells in both cases, although their existence in shoot apices has been seriously challenged. The initials of the cambium, however, do not correspond in their activity to the classical concept of apical initials as represented by the apical cells of lower vascular plants. On the other hand, the concept of apical initials and their function in higher plants is very poorly defined and is particularly vague in regard to patterns of segmentation. It is possible that close attention to the pattern of segmentation in the cambium could provide a better understanding, or at least a more precise theory, of the functioning of initials in the shoot apex (Newman, 1956).

One of the particularly striking features of the vascular cambium is the close correspondence in structural organization between the meristem and the tissues that it produces, the secondary xylem and phloem. This is in marked contrast to the terminal meristem, particularly that of the shoot, in which the meristem has its own distinctive organization quite apart from the initial differentiation of tissues. This, of course, is correlated with the fact that the cambium produces certain tissues only, whereas the terminal meristems initiate an entire shoot or root. There seems to be a question here of levels of differentiation. The shoot and root apices, as has been pointed out, are differentiated in the embryo as shoot- and root-producing centers and are seemingly stabilized at this level, retaining for prolonged periods the capacity for cell production while the derivatives they produce undergo maturation. It may be suggested that something of a similar sort occurs in the process of cambial initiation, but at a relatively advanced stage of vascular tissue differentiation, the procambial stage. The resulting meristem is thus very strongly committed to the initiation of vascular tissues, while retaining the capacity for indefinite cell production. It is, in one sense, a histogen, although probably not in the sense in which this term has been used for shoot and root apices. Whether the mechanism of stabilization is the same, or even comparable, in the two cases is, of course, another way of posing the question with which this general comment began.

The close correspondence between the cambium and the mature tissues it produces greatly facilitates the study of this meristem because a record of the changes that occur in the cambium with the passage of time is left behind, particularly in the relatively permanent secondary xylem.

This record reveals that the cambium is a surprisingly dynamic population of meristematic cells and one in which the changing patterns of cell division and cell elongation suggest a rather precise control. At present there is very little information about the factors that regulate the frequency and the orientation of cell divisions and the extent and direction of cell elongation but a limited amount of experimental work has begun to point out ways in which controlling mechanisms may be explored. This is the subject of the following chapter.

REFERENCES

Bailey, I. W. 1954. Contributions to plant anatomy. Chronica Botanica. Waltham, Mass.

Bannan, M. W. 1955. The vascular cambium and radial growth in *Thuja occidentalis* L. Can. J. Botany **33**:113–138.

———— 1956. Some aspects of the elongation of fusiform cambial cells in *Thuja occidentalis* L. Can. J. Botany **34**:175–196.

———— 1957a. Girth increase in white cedar stems of irregular form. Can. J. Botany **35**:425–434.

———— 1957b. The relative frequency of the different types of anticlinal divisions in conifer cambium. Can. J. Botany **35**:875–884.

———— 1960. Ontogenetic trends in conifer cambium with respect to frequency of anticlinal division and cell length. Can. J. Botany **38**:795–802.

Barghoorn, E. S., Jr. 1940a. Origin and development of the uniseriate ray in the Coniferae. Bull. Torrey Bot. Club **67**:303–328.

———— 1940b. The ontogenetic development and phylogenetic specialization of rays in the xylem of dicotyledons. I. The primitive ray structure. Am. J. Botany **27**:918–928.

———— 1941. The ontogenetic development and phylogenetic specialization of rays in the xylem of dicotyledons. II. Modification of the multiseriate and uniseriate rays. Am. J. Botany **28**:273–282.

Barghoom, E. S. 1964. Evolution of cambium in geologic time. *In* M. Zimmermann [ed.], The formation of wood in forest trees. pp. 3–17. Academic Press, New York.

Esau, K. 1948. Phloem structure in the grapevine and its seasonal changes. Hilgardia **18**:217–296.

Newman, I. V. 1956. Pattern in meristems of vascular plants. I. Cell partition in living apices and in the cambial zone in relation to the concepts of initial cells and apical cells. Phytomorphology **6**:1–19.

Wilson, B. F. 1964. A model for cell production by the cambium of conifers. *In* M. Zimmermann [ed.], The formation of wood in forest trees. pp. 19–36. Academic Press, New York.

———— 1966. Mitotic activity in the cambial zone of *Pinus strobus*. Am. J. Botany **53**:364–372.

SIXTEEN

Secondary Growth—

Experimental Studies

on the Cambium

The fact that the activity of the vascular cambium can be traced very precisely in its mature derivatives suggests that this meristem ought to be a useful system in which to study the control of developmental phenomena. On the other hand, its relative inaccessibility to direct manipulation and to observation are obstacles to the kind of experimentation that has been so profitable in the case of the shoot apex. It is perhaps not surprising, therefore, that there has been very little experimental work dealing with the control of developmental patterns, while at the same time such phenomena as seasonal activation have received considerable attention. There has also been a great deal of interest in the participation of the cambium in wound reactions and healing and in the establishment of tissue unions in grafts of various types, both areas of considerable applied significance. The relatively few studies that have dealt with fundamental aspects of development in the cambium together with information that may be extracted from a number of applied investigations give strong indications that the experimental approach could be as vaulable a tool in the understanding of cambial problems as it has been in the case of the terminal meristems.

Cambium cultures

One of the most effective techniques employed in evaluating the potentialities and the regulation of the terminal meristems of shoot and root has been the excision of these centers with a minimum of their differentiated derivatives and their culture on controlled nutrient medium. The application of this method to the cambium—although beset with inherent difficulties because only small pieces can be excised and because removal involves extensive injury—should, if successful, answer the same kinds of questions concerning the degree of developmental autonomy of the meristem as opposed to external controls as were investigated in the case of terminal meristems. Interestingly enough, the fragmentation and injury involved in the removal of pieces of cambium, including the initials and their most recent derivatives, do not appear to be fatal to these cells. Bailey and Zirkle (1931) have shown that tissue slices cut from the cambial zone of diverse conifers and dicotyledons with a sledge microtome and containing cambial initials and recent derivatives may be kept alive and under intermittent observation for periods exceeding two months in a medium consisting solely of tap water. The vitality of the contained cells was revealed by cytoplasmic streaming and by responsiveness to treatments, particularly, in these experiments, changes in external pH. The cells, however, did not undergo division under these conditions, and the addition of various nutrients such as mineral salts or sucrose or the substitution of distilled water merely reduced the duration of viability of the explants.

The absence of cell division in these cultures does, however, seem to be a matter of the lack of the proper nutrients and growth factors. It is now well known that tissue slices including the cambial zone, if removed aseptically from a large number and variety of species and explanted to a nutrient medium containing mineral salts, sugar, certain vitamins and an auxin, will proliferate freely and indefinitely. In fact, historically (Gautheret, 1942), such tissues were among the first from which true plant tissue cultures were obtained, that is, cultures capable of unlimited growth. One of the most interesting characteristics of these cultures is the fact that the cambium, although its cells divide freely, does not continue divisions according to the pattern in the intact plant and does not initiate xylem and phloem. Rather, all the cells capable of division segment into small, approximately isodiametric cells, and the result is a callus culture consisting of relatively homogeneous parenchyma in which the cambium loses its identity. The only cambium ever present in callus cultures is one that may differentiate later. Thus a portion of the cambium, even if excised with a considerable number of its derivatives, is unable to maintain a normal pattern of division and differentiation when isolated from the plant.

The importance of pressure

The difficulty with isolated pieces of cambium, then, is not that they cannot undergo cell division but rather that the cambial layer cannot maintain the orderly division pattern so essential to its organization. Not surprisingly there has been a great deal of speculation about the possible nature of the factors that are lacking in the isolated pieces, or in other words, the factors that control orderly division patterns in the intact plant. Experiments by Brown and Sax (1962) provide strong support for the long suspected conclusion that the major factors in this phenomenon may be physical in nature rather than chemical. These workers investigated cambial development in partially isolated strips of bark containing cambium which, since they remained attached to the plant at one end, could presumably obtain all substances necessary for their continued development. In *Populus trichocarpa* and *Pinus strobus* bark strips eight centimeters long and one and a half centimeters wide, severed at their basal ends but left attached at the tops, were pulled away from the trunk and encased in polyethylene bags (Fig. 16.1a). Following this, the enclosed strips either were left free or were reinserted into the slots from which they had been separated with varying amounts of pressure applied. The separation of the strips during a period of intense cambial activity was shown to occur in the zone of recently formed and as yet incompletely differentiated secondary xylem elements so that the cambium was intact within the strip.

In the free strips there was a very rapid proliferation on the inner face primarily by the immature ray cells but also to some extent in other immature xylary elements which led to the establishment of a layer of callus up to 5 millimeters in thickness by the end of three or four weeks. During this callus development the cambium itself continued to proliferate in a more or less orderly fashion, but its early xylary derivates were shorter than normal and somewhat irregular in shape. Whether this indicates abnormalities in the cambial initials or in their derivatives is not clear; but because the formation of relatively normal xylem elements is soon reestablished, it would seem that the cambial initials were little affected. After the callus pad had reached the dimensions just noted, a periderm formed near its surface and beneath this cambial activity, which appeared to be quite normal, gradually extended across the pad from the margins in continuity with the original or outer cambium of the strip. Cross sections of such a bark strip at this time reveal a complete ring of cambium with its associated secondary vascular tissues (Fig. 16.1b).

In a parallel series of plants, the separated bark strips, encased in polyethylene, were forced back into the slots on the trunks and either bound with an elastic grafting band or maintained under measured pressure by the application of a mercury column of a given height with a basal reservoir having one flexible face appressed to the bark strip. The development of the bark strip was strikingly different from those that developed without pressure (Fig. 16.1c). In the first few days callus began

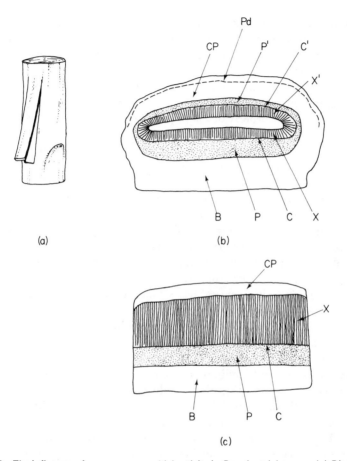

Fig. 16.1 The influence of pressure on cambial activity in *Populus trichocarpa*. (a) Diagram showing the manner in which a bark strip was cut and lifted in the experiments of Brown and Sax, 1962. (b) Organization of the tissues that differentiated in a bark strip that was left free for sixty days. (c) Organization of differentiated tissues in a bark strip that was reinserted and maintained under pressure for sixty days. Key: B, bark; C, original cambium; C', new cambium; CP, callus pad; P, phloem formed by original cambium; P', phloem formed by new cambium; Pd, newly formed periderm; X, xylem produced by original cambium; X', xylem produced by new cambium. In *b* and *c* the original outer surface of the bark strip is to the bottom of the diagram. (Adapted from C. L. Brown and K. Sax, 1962. Amer. J. Bot. **49**: 683.)

to form as in the previous case, but this continued only until irregularities between the two opposing surfaces had been filled. At this time the callus cells differentiated as lignified elements of the tracheary type. The cambium itself functioned quite normally throughout this period and the immature xylem elements present at the beginning of the experiment differentiated in a nearly normal fashion. The range of pressures applied to the strip in this experiment was 0.25 to 1.0 atmosphere, but all were equally effective in promoting normal development within the strip, as was the grafting rubber band.

This experiment has been interpreted by Brown and Sax as demonstrating the important role of physical pressure, or perhaps confinement, in maintaining the orderly development of the cambium and its derivative tissues. The absence of pressure in the free strips results in the extensive proliferation of recent cambial derivatives, whereas when pressure is applied this process goes on only until all spaces are filled and, presumably, an even pressure results. In the isolated strip, when the callus pad has developed to an extent sufficient to exert physical constraint on the inner tissues, development becomes relatively normal and, in fact, a new cambium is formed in the callus itself. The process of callus formation described in this experiment in which the proliferating tissue originates in cambial derivatives, notably immature or even mature ray cells, is regarded as being the usual reaction to wounding in woody species, and the cambial initials themselves seldom if ever play an important role (Sharples and Gunnery, 1933).

It is now pertinent, with this background information, to return to the question of behavior of cambial explants in nutrient culture. Brown (1964) has investigated the development of such explants from *Populus deltoides* on a nutrient medium adequate to support proliferation. On this medium explants consisting of inner phloem and the cambial zone showed extensive proliferation which involved parenchyma cells throughout the explant and the cambial initials themselves, resulting in a complete loss of identity of the cambium. Thus, although the proliferation was of the same type observed in the free bark strips in the tree experiments, it was considerably more extensive, reflecting either the complete isolation from the plant or the fact that the thinner tissue slice used in culture may provide less restraint or a more generalized distribution of nutrients. It is therefore of particular interest to note that when such explants were subjected to direct pressures of 0.1 or 0.05 atmosphere by means of a mercury column and basal rubber well the development of callus was greatly reduced to a few cell layers in irregular surface depressions and the cambial initials retained their identity. Moreover there was evidence that the cambial initials had given rise to derivatives that differentiated normally. Thus there is very strong evidence for the important role of physical forces, presumably resulting in the intact plant from the constraint imposed by outer tissues and the pressure generated by an expanding inner core, in maintaining the pattern of orderly divisions in the cambium and the normal processes of differentiation. In this connection it may be recalled that the constraint imposed by the archegonial venter upon the zygote and early embryo has been implicated in the regulation of orderly divisions in early stages of fern development.

Orientation in the cambium

The descriptions presented in the previous chapter have shown that the cambium consists of cells and cell groupings that are highly oriented. The elongate fusiform initials and the fusiform groups of ray initials lie with their long axes parallel to one another and generally parallel to the long axis of the organ in which they are found. The growth activities of this meristem are intimately connected with this orientation because it has been shown that cell divisions are precisely oriented in these cells. The role of physical pressure in maintaining orderly patterns of cell division in the cambium has been indicated; but it would seem to be desirable to consider the extent to which this or other factors influence the orientation of cells and cell groups. Several possibilities immediately suggest themselves for consideration. The orientation of cambial initials might be relatively labile, being imposed by the polarity of the axis in which they are located, or more specifically by the polarized movement of nutrients, water, or hormones along the axis. Similarly one might imagine that the orientation is established when the cambium is initiated and is then maintained by the influence of mature derivatives which, of course, have the same orientation. Because these derivatives are also the conducting cells of the axis, this kind of regulation would be difficult to distinguish from the previous suggestion. Alternatively the cambial initials could be to a large extent insensitive to stimuli external to themselves in the matter of orientation.

In attempting to evaluate these possibilities, one is faced with a large body of literature dealing with observations incidental to the study of grafting and wound healing, which dates back to the early nineteenth century. Much of this information is difficult to evaluate in terms of the behavior of cambial initials, but collectively it provides good evidence for the orientation of cambial initials along the lines of major transport in the axis and for their change in orientation when the direction of transport is altered. In some cases the orientations are extremely complex. One well-documented example of this kind of result illustrates the evidence upon which this conclusion is based. MacDaniels and Curtis (1930) carried out experiments on young apple trees in which helical strips of bark including the cambium which extended twice around the trunk were removed and the gap was filled with wax to prevent regeneration (Fig. 16.2). This operation left a continuous helical bark strip several inches in width connecting the upper and lower intact regions of the trunk, its phloem providing the sole connection between the two for the

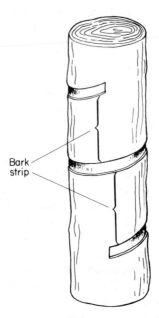

Fig. 16.2   Diagram showing the technique used by MacDaniels and Curtis, 1930, to produce spiral ringing of young apple trees.

conduction of organic nutrients. In this strip the direction of movement of both nutrients and water must have been changed by about 45 degrees, the pitch of the helix. The first few layers of xylem elements that differentiated after the operation were vertically-oriented, but they departed from normal in that the perforation plates were on the lateral walls rather than on the end walls. Progressively, however, the orientation of the elements changed, presumably indicating a change in orientation of the cambial initials, until the long axis of the tracheary cells and of the rays was parallel to the axis of the helical strip, that is, about 45 degrees from the vertical. The elements of the newly formed phloem were correspondingly changed in their orientation, further indication that the cambium itself, which was not examined, had undergone a change in orientation of its intials.

Although the change in orientation that resulted in this experiment certainly had the effect of improving transport of materials through the newly formed xylem and phloem, there is no direct proof that transport of materials was the causal factor involved in the shift, or indication of which materials might be considered most important in producing the change. However, because even in a completely ringed tree water can be conducted through the xylem that remains, it would seem most likely that materials transported in the phloem would be implicated.

Several interesting points arise from this experiment. First, it is clear that although the cambium reoriented within the first season, its reaction was not immediate. This was not because of a delay in the appearance of the stimulus to reorientation because an immediate effect upon the differentiation of tracheary cells was noted. Rather, the initials of the cambium must require some time to effect the change, or perhaps they tend to resist it. This raises the question of how the reorientation is accomplished. The cambial cells might subdivide into a sort of callus within which a new cambium could form in the new orientation, but careful examination revealed no evidence for this. Thus the cells must have shifted their axes. By referring back to the earlier discussion of growth patterns in the cambium, it is found that the only apparent mechanism by which this could be accomplished would be apical intrusive growth following pseudotransverse or oblique anticlinal divisions. Possibly this process is accelerated in the conditions of the experiment, and the elongations following division occur along the new axis. This process, if it occurs, could be detected in the secondary xylem formed during the period of transition, but this has not been examined to date. An interesting further experiment could be performed by carrying out the same operation on a species having a storied cambium in which there is no apical intrusive growth. Possibly the cambium of such a species would be unable to respond, unless under abnormal conditions the process does occur. The reorientation of the rays in any case is more difficult to explain, and a study of this process would be most interesting.

There is, then, substantial evidence to support the idea that cambial initials tend to be oriented along the lines of major transport in the axis; but at the same time experimental evidence of another type suggests that the cambial initials are relatively autonomous in this regard. Thair (1968) has investigated this question in several species including *Malus spp.* (ornamental crabapples), *Thuja occidentalis*, *Cornus stolonifera*, and *Sorbus aucuparia* by removing small, square patches of bark and replacing them, or exchanging them with comparable pieces from other trees, at orientations of 90 or 180 degrees (Fig. 16.3a). As in experiments mentioned earlier, when bark is lifted the separation occurs in the region of immature xylem elements so that the cambium is removed with the bark piece and is rotated with it. Perhaps the greatest interest in these experiments is attached to the pieces that were rotated 90 degrees so that the cambial initials lay at right angles to the axis of the stem and to the major direction of transport. In this case the rather surprising result was that the rotated cambium continued to function normally in the altered orienta-

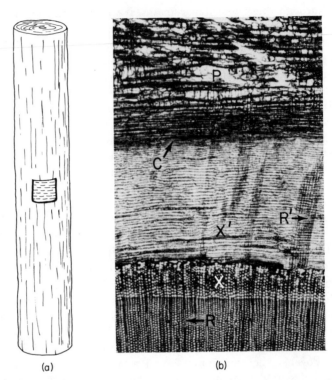

(a)   (b)

Fig. 16.3 Experimental rotation of the cambium. (a) Rotation of a square patch of bark through 90°. (b) Transverse section of a stem cut through such a patch in *Thuja occidentalis* (white cedar) one hundred days after the rotation operation. Key: C, cambium; P, phloem; R, ray in original xylem; R′, ray in new xylem; X, original xylem; X′, xylem formed after the operation. (Photograph courtesy B. W. Thair.)

tion and that secondary xylem and phloem were formed in a small block oriented at right angles to that of the host plant (Fig. 16.3b). In these experiments, of course, there was no question of releasing pressure because this was supplied by a grafting rubber band until the graft union had formed about two weeks after the operation was made. What is important here is that the cambial initials did not undergo reorientation; and in certain cases in which the experiment was allowed to continue, this development was maintained through several growing seasons as revealed by the development of growth rings in the secondary xylem matching those of the host plant. Moreover, Thair has shown that the same result can be obtained with circular patches of bark oriented at angles of 30-degree intervals from 0 to 180 degrees. With almost incredible accuracy the cambium retained the orientation given to it and laid down secondary conducting elements in this orientation.

These results argue forcefully for the conclusion that the orientation of initials in the cambium is insensitive to external influences, but several additional observations argue against this extreme view. In the same trees in which the grafted cambium retained its new orientation independently of the direction of transport in the axis, the cambium of the host or stock showed considerable reorientation in the vicinity of the graft, with the result that the graft was largely bypassed in the transport system. Thus reorientation was found in the same experiments, but under different circumstances. Secondly, various abnormalities were noted in the secondary xylem elements formed in the period immediately after the operations were made, notably the occurrence of perforation plates on the side walls. Unfortunately the phloem was not examined in the same detail so that it is not known whether comparable modifications occurred there as well. It does appear, however, that the problems created by an orientation at right angles to the lines of transport had an effect upon the graft, although not necessarily directly upon the cambium. However, before the end of the first growing season these abnormalities ceased, although there was no reorientation. It was then discovered that both in the square and in the circular grafts, although all of the edges of the grafts had united with the stock by the formation of callus, only at the original top and bottom of the grafted pieces had a true continuity been established in the cambium with a resulting continuity in derivative xylem and phloem. Furthermore it appeared that the original upper ends of grafted cambial initials had made contact with the bases of stock initials and the bases of graft initials with the tops of stock initials with the result that, in the case of a 90-degree graft, for example, it was possible to tell by later examination in which direction the pieces had been rotated. Thus it appears that the grafted pieces were incorporated into the transport system in a special way such that they were supplied with materials and were not subsequently subjected to the kind of stress which, in other cases, has resulted in reorientation. It also appears that the cells of the cambium have an inherent polarity that causes them to make the correct end-to-end contacts in graft unions.

General comment

In attempting to make some general statement about the control of developmental processes in the cambium, it is evident that the major problem is a lack of information. Of the numerous experiments that suggest themselves as obvious necessities, so few have been done that one can do little else than talk about problems for future research. At its origin the cambium is endowed with a highly oriented organization. Perhaps the most important aspect of this orientation is that relating to planes of cell division. Experiments have shown rather clearly that physical pressure, of the type to which the cambium is normally exposed in the intact plant, is essential for the maintenance of an orderly division pattern; but there is nothing which indicates that the pressure is responsible for the pattern. It is tempting to speculate that differences in physical forces might play a role, for example, in regulating the frequency of anticlinal divisions that bring about an increase in circumference; but there is nothing factual to indicate that this is the case.

As for the orientation, this seems to be imposed upon the cambium at its origin and, as the grafting experiments have shown, can be maintained by pieces of cambium regardless of orientation relative to the axis of the organ in which the cambium is found. The individual cambial initials give evidence of possessing an inherent polarity that is independent of the surroundings. In other words, the cambium can be relatively autonomous in maintaining its orientation and the functional features that accompany it. However, there is also evidence that the cambium is sensitive to the lines of flow of nutrients in the axis, and from the point of view of utility it would seem logical that it should be. If a particular unit of cambium is not subject to stress, it can retain indefinitely any orientation that it is given; but if it is under stress it can assume an orientation parallel to the lines of flow. It is not dependent upon external factors for the maintenance of its orientation but it does respond to them. There are, of course, many unanswered questions here. What, for example, is the real nature of the stress and how is it applied to the cambium? How does the cambium achieve a reorientation in those cases in which it occurs? These questions and many others that are obvious reveal the cambium as an extremely promising object for future experimental studies on development.

REFERENCES

Bailey, I. W., and C. Zirkle. 1931. The cambium and its derivative tissues. VI. The effects of hydrogen ion concentration in vital staining. J. Gen. Physiol. **14**:363–383.

Brown, C. L. 1964. The influence of external pressure on the differentiation of cells and tissues cultured *in vitro*. *In* M. Zimmerman [ed.], The formation of wood in forest trees. pp. 389–404. Academic Press, New York.

Brown, C. L., and K. Sax. 1962. The influence of pressure on the differentiation of secondary tissues. Am. J. Botany **49**:683–691.

Gautheret, R. J. 1942. Manuel technique de culture de tissus vegetaux. Masson et Cie, Paris.

MacDaniels, L. H., and O. F. Curtis. 1930. The effect of spiral ringing on solute translocation and the structure of the regenerated tissues of the apple. Cornell Univ. Ag. Expt. Station Memoir #133.

Sharples, A., and H. Gunnery. 1933. Callus formation in *Hibiscus rosa-sinensis* L. and *Hevea brasiliensis* Müll. Arg. Ann. Bot. **47**:827–839.

Thair, B. W. 1968. Cambial polarity as revealed by grafting experiments. Thesis, University of Saskatchewan.

SEVENTEEN

The Cellular Basis

of Organization

In the development of the plant body, one cell, the zygote, is able to express the full genetic potentialities of the organism, that is, it normally gives rise to the whole plant. All other cells express their potentialities less completely, and such expression is progressively restricted as one proceeds through the later stages of tissue and cell differentiation. Ultimately an individual cell differentiates as a highly specific entity such as a tracheid or a sieve element, which clearly expresses only a small portion of the total genetic capacity of the organism. Where differentiation brings about a drastic change in the morphology of the cell, such as the death of the protoplast in a tracheid, or the loss of the nucleus in a sieve element, no further expression of genetic potentiality is possible and differentiation is irreversible. In most cells, however, no such irrevocable loss occurs; and if differentiation is a manifestation of selective gene action rather than of mutational changes in the nucleus, as was discussed in Chapter Thirteen, then reactivation of these cells by the appropriate stimuli might be expected to result in further and perhaps different expressions of their potentialities. In fact, this expectation is amply realized under many conditions, both natural and artificial, in which differentiated plant cells undergo further development and realize different and more complete expressions of their potentialities than in their original differentiation.

Developmental potentialities
of isolated cells

Proliferation of single cells

Around the turn of the century, the German botanist Gottlieb Haberlandt (1902) expressed a conviction that isolated plant cells, if properly nurtured, ought to be capable of developing into whole plants as the zygote does. Although Haberlandt's own experiments failed to achieve this objective as did those of all other investigators for more than 50 years, modern tissue culture methods have vindicated his view that the zygote is not unique in its ability to give rise to an entire plant. In order to test the potentialities of single cells it was necessary first to culture them under conditions in which they would divide repeatedly; and this proved to be a refractory technical problem. Repeated attempts to isolate single cells from callus tissue and to culture them on media that supported the growth of larger tissue masses failed even when the cells were placed in small volumes of medium such as in hanging drops.

The successful solution of this problem was provided by Muir et al. (1958) working with callus tissue cultures of *Nicotiana tabacum* (tobacco) and several other species. Single cells were carefully picked from the surface of friable tissue masses or were removed from liquid cultures in which tissue dissociation had occurred and were placed separately on pieces of filter paper resting on the surface of established cultures of the same or a different species (Fig. 17.1). This method was adopted in the expectation that the established culture might provide substances necessary to initiate division in the isolated cell and that would be lacking in an otherwise complete nutrient medium. Under these conditions single cells divided repeatedly and produced callus masses that could then be subcultured indefinitely. Thus tissue clones were established that had their origin as single cells. Careful examination of the filter paper barrier between the new and the established cultures supported the conclusion that the tissue masses arose from the single cells, because there was no evidence that cells of the established culture had grown through or around the filter paper. Where the established culture and the single cell clone were from different species, any such invasion through the filter paper would, of course, have been obvious, but none was observed.

Although this method established the ability of a single cell to proliferate, it did not permit microscopic observation of the stages of development. For this purpose attempts have been made to establish isolated cells in microculture. There was success (Jones et al., 1960) in obtaining

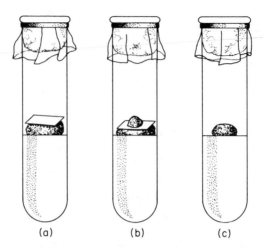

Fig. 17.1   Diagrams of the procedure used by Muir, 1958, to establish tissue clones from single cells. (a) A single cell taken from the surface of a piece of callus tissue or from a liquid suspension culture is placed on filter paper lying on the top of a large tissue culture. (b) Single cells proliferate to produce macroscopically visible masses of tissue. (c) Growing tissue masses are removed from the filter papers to separate culture. (Based on data in W. H. Muir et al., 1958. Amer. J. Bot. **45** : 589.)

cultures from single cells of *Nicotiana tabacum* isolated in small drops of a nutrient medium that had been conditioned by prior growth of the callus. The cells were cultured on microscope slides, and the early stages of proliferation could be observed microscopically. Subsequently Vasil and Hildebrandt (1965a) eliminated the necessity for conditioning the medium by selecting only relatively small cells from liquid shake cultures at the time of maximum tissue dissociation and by slightly altering the composition of the medium. The divisions of the isolated cell and its derivatives could be followed until a cluster of cells had been produced, which was capable of continued growth. Thus it was demonstrated by these experiments that an isolated cell can proliferate apart from the influence of other cells.

Growth and morphogenesis
in suspension cultures

While these studies were being carried out, Steward and his coworkers (1958) were developing a different approach to the study of potency of isolated cells (Fig. 17.2). These workers explanted small pieces of secondary phloem tissue from the root of *Daucus carota* (carrot) and cultured them in large rotating flasks of liquid medium. The tissues

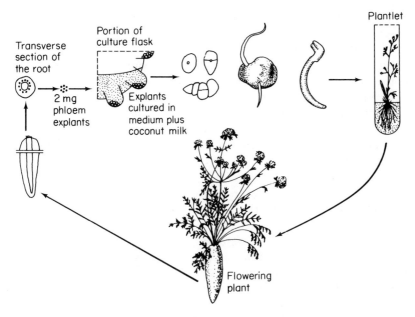

Fig. 17.2 Development of carrot plants from root cells in tissue culture. The diagram shows the technique used by Steward, 1964. to obtain carrot plants. A slice of carrot tap root, shown at the top left, is used to obtain explants of tissue. Phloem explants cut from the disc are inoculated into liquid culture medium in specially constructed flasks that rotate slowly. Single cells and cell clumps separate from the tissue masses and form a suspension. Some of the varied single cells and clumps are shown in the middle of the top row of the diagram. Some of the small clumps produce roots. Rooted nodules are transferred to agar-hardened medium where they produce shoots. When transferred to soil these plants form a typical tap root and produced flowers. Tissue from the tap root can be used to repeat the cycle. (Adapted from F. C. Steward et al., 1964. Science **143**: 20.)

proliferated rapidly and in the constantly agitated medium numerous individual cells and groups of cells separated from the developing callus masses and remained freely suspended in the liquid medium, producing a suspension culture. Such suspensions could be maintained through repeated transfers by inoculation of fresh culture medium with small amounts of material from a previous culture. Microscopic examination of the suspension revealed the presence of numerous single cells and clusters ranging from two to many cells. Although under these conditions the proliferation of individual cells could not be followed, careful study of the range of cells and cell aggregates led these workers to the conclusion that isolated cells do divide, following a variety of patterns, and produce small nodules of callus tissue.

A surprising development in the continuing suspension cultures of carrot root tissue was that many of the nodules that presumably had arisen from single cells gave rise to roots. As long as the nodules remained in suspension culture, although roots continued to grow, no other organized structures appeared. When, however, rooted nodules were transferred to a medium solidified with agar so that they were maintained in a fixed position, many also gave rise to shoots. The result, therefore, was the production of complete plantlets which, as they grew, developed tap roots characteristic of the carrot plant. Ultimately some of these were transferred to soil where they completed the life cycle by flowering and producing seeds.

The general conclusion from these studies is that isolated plant cells other than the zygote are able to express the full genetic potentialities of the organism by giving rise to entire plants. Although the immediate source of these plants was cells derived from a suspension culture, it is significant that the cultures were originally established from differentiated tissues of the plant. However, in the suspension culture method it is not possible to follow development of an individual cell from its first division to the establishment of a shoot and root. The conclusion that single cells were the source of the plantlets was derived from study of numerous entities in different stages of development and the deduction that these revealed the steps through which an individual entity proceeds. Although this conclusion seems justified on the basis of the evidence presented, it is reassuring to have it confirmed in detail. Vasil and Hildebrandt (1965b), in a continuation of their earlier studies on tobacco cells grown in microculture, found that if the isolated cells were taken from recently established tissue cultures the calluses that resulted from their proliferation gave rise to roots and shoots and ultimately to plantlets as in the case of the nodules of carrot tissue from suspension culture (Fig. 17.3). As with carrot, these plants could be transplanted to soil where they flowered.

Although these studies demonstrate that individual cells isolated from callus cultures are capable of fulfilling the role normally reserved for the zygote—that is, the production of a whole plant—the fact remains that the means by which this is accomplished are different from those exhibited in zygote development. The initiation of shoot and root apices in a small nodule of callus derived from a single cell is not the same as the direct formation of an organized embryo from a zygote. Considerable interest is therefore attached to several reports of the production of embryo-like structures, or *embryoids*, in suspension cultures (Fig. 17.4). Steward et al. (1964) have described the development of large numbers of embryoids, involving a high proportion of the suspended cells and cell aggregates, in suspension cultures derived originally from embryos of carrot. Very similar developments were observed by Halperin and Wetherell (1964)

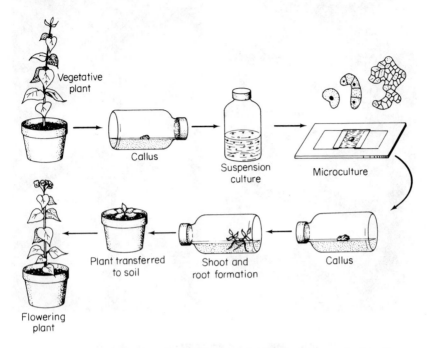

Fig. 17.3   Development of tobacco plants from single tissue culture cells. The diagram shows the technique used by Vasil and Hildebrandt, 1965. A hybrid vegetative plant of the cross *Nicotiana glutinosa* ×*N. tabacum* is used as the source of pith cells which are transferred aseptically to establish a callus culture on an agar-hardened medium. Transfer of the tissue to liquid medium results in development of a suspension culture. Single cells removed from the suspension culture to microslide culture can be observed directly or by time-lapse photography as they divide and form small tissue masses. These tissue masses are removed to agar medium when they are large enough to be manipulated. On the agar medium shoots and roots are initiated. Plants are then transferred to soil where they develop vegetatively, flower, and set seed. (Adapted from V. Vasil and A. C. Hildebrandt, 1965. Science **150**: 889.)

in cultures initiated from root or petiole tissue of the wild carrot, which is, in fact, the same species as cultivated carrot. The formation of large numbers of embryoids in wild carrot suspension culture has facilitated the investigation of developmental sequences, although of course it is not possible in such a system to trace the development of individual entities. Halperin (1966) has compared the development of embryoids with that of normal seed embryos of the same species and has shown that the later stages of embryoid growth, from the globular stage or the heart-shaped stage, are remarkably like those of normal embryogeny. There appear to be some minor but characteristic departures in embryoids from the normal pattern, such as larger size at comparable stages, a delay in the

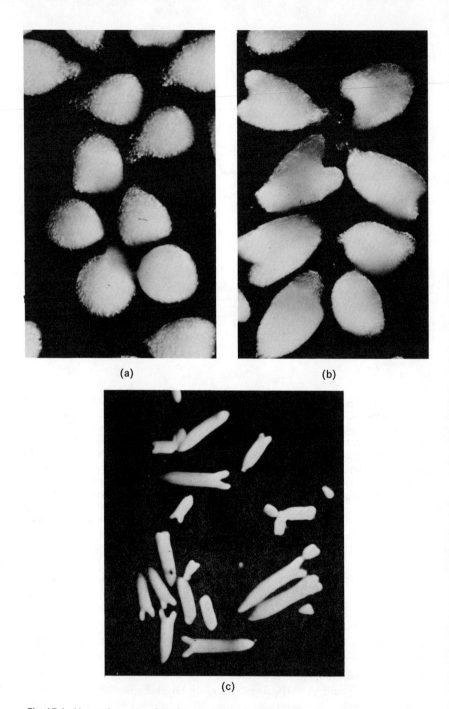

(a)

(b)

(c)

Fig. 17.4   Vegetative embryoids of carrot differentiated in suspension culture. (a) Embryoids at globular and early heart stages. (b) Heart stage embryoids. (c) Maturing embryoids. (a, b) ×107, (c) ×40. (W. Halperin, 1966. Amer. J. Bot. **53**: 443.)

initiation of cotyledons and a precocious vacuolation of the future parenchymatous tissues; but the end point, prior to germination, is a cylindrical, dicotyledonous embryo with well-organized shoot and root apices
and the three tissue systems delimited as in seed embryos. Examination
of the earliest phases of embryoid development, however, reveals more
extreme differences from the normal pattern. Globular embryos arise as
three-dimensional, bud-like developments at the surface of small cell
aggregates that are without apparent organization, and more than one
may arise from a single aggregate (Fig. 17.5). These primordia develop

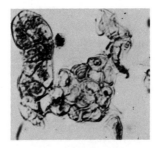

Fig. 17.5 Early stage in the differentiation of a carrot embryoid from cells in suspension
culture. The globular proembryoid consists of small, densely cytoplasmic cells attached to
vacuolated cells that resemble the suspensor of a sexually produced embryo. ×360. (W.
Halperin, 1966. Amer. J. Bot. **53**: 443.)

directly into globular embryos, and the cells of the original aggregate may
remain attached to them as a suspensor-like appendage. This pattern of
early development contrasts sharply with the normal embryogeny of
carrot in which the zygote gives rise to an elongated filament from the
terminal few cells of which the globular stage arises, the remainder of the
filament producing the suspensor. Interestingly enough, the embryoid
pattern described by Halperin is paralleled in a few species of angiosperms
in which the zygote develops into an embryogenic mass from which
several embryos arise by budding. The differences in pattern that occur
between embryoids and seed embryos in the early stages of development
should not, however, be allowed to obscure the main result of these studies,
namely that the overall sequence of development is remarkably similar
to that of a zygote expressing its potentialities in an embryo sac. This is
reinforced by the fact that the early division patterns in embryos of many
species including carrot are known to be variable.

The inescapable conclusion drawn from the experimental studies that have been outlined is that differentiated plant cells retain their full genetic potenialities, and under the appropriate conditions can express them, if not directly, at least indirectly after cell proliferation has occurred. Thus the question that stimulated this investigation, "Do differentiated cells retain their full potentialities and can they express them?" has now been replaced by another question, "Since differentiated cells retain their full potentialities and can express them, under what conditions is this new expression possible?" This question is raised by the fact that in the normal course of events, differentiated cells within the plant body do not give rise to new plants and, in fact, would preclude the existence of an organized plant if they did. It is to this question that attention must now be turned.

Requirements for embryoid formation

When entire plants were first produced in suspension cultures presumably from single cells, and in microcultures demonstrably from single cells, it was assumed by many that isolation of the cell was essential for the expression of its latent zygotic potentialities. In suspension culture, of course, the cells are not truly isolated from the chemical influences of other cells but they are removed from any physical contact or restraint to an even greater extent than is the attached zygote in the embryo sac. Later investigators, however, have cast serious doubt upon this assumption even to the extent that it is now suggested that the formation of embryoids occurs more readily in cell clusters separated from a callus than it does from the isolated cells of a suspension culture. A number of reports have appeared in which typical embryoids are described as arising at the surface of coherent tissues, where the cells clearly are not physically isolated. Even in carrot, callus cultures can produce numerous embryoids on their surfaces without dissociation of the tissue. One of the most interesting examples of this phenomenon was described by Konar and Nataraja (1965) who observed the initiation of typical embryoids from the epidermis of the stem of *Ranunculus sceleratus* seedlings growing in sterile culture. Clearly, then, a cell need not be physically isolated before it can function as a zygote; in fact, there can be no requirement that the tissue of which it is a part be without organization.

The original suspension cultures in which embryoids and other manifestations of totipotency were observed were established in nutrient media that included the liquid endosperm coconut milk. It is not surprising that this rich nutrient broth which normally nurtures the development of embryos should have come to be regarded as an essential, or at least optimal, medium for the expression of cellular totipotency. In fact, other similar materials such as the female gametophyte of *Ginkgo* and the liquid endosperm of horse chestnut and several other species were ascribed corresponding significance. It was to be expected that considerable effort would be devoted to attempts to determine exactly what coconut milk and similar materials from other species contain that might be essential for the expression of cellular totipotency. On the other hand, there is increasing evidence that, at least in carrot, embryoids are abundantly initiated in the absence of coconut milk or any other such natural product (Reinert, 1967). In fact, coconut milk appears to be somewhat antagonistic to the development of these organized structures. Also it has been shown that reduced nitrogen compounds, either organic such as amino acids, or inorganic in the form of ammonium salts, greatly enhance embryoid formation in carrot. Significantly, coconut milk is a rich source of reduced nitrogen, particularly in the form of amino acids. There are indications that carrot tissue requires a low concentration of auxin in its medium as a prerequisite to embryoid development. This, too, is a component of coconut milk. Coconut milk is also known to contain substances having cytokinin properties, that is, substances that promote cell division. Undoubtedly this component is of significance in initiating cell division in explanted mature tissues, and perhaps in maintaining it actively in the resulting cultures. There is no evidence, however, that such substances are required for the organization of embryoids.

In considering the conditions under which cellular totipotency can be expressed, it is important to remember that, as in so many other developmental phenomena, differences among species can be striking. Although embryoid formation has been described in a number of species, there are others that have failed to respond to any treatment. For example, Halperin and Wetherell, at the time of their success with wild carrot, failed to obtain any organized development in cultures derived from five other species of the same family (Umbelliferae) treated in the same way. Similarly Steward failed to obtain results with tissues of several other species comparable to those obtained with carrot cultures.

Clearly the conditions that permit totipotency to be expressed by differentiated cells are difficult to define precisely and are almost certainly diverse. It does not appear that any specific substance or group of substances is universely required. The cells must be induced to divide, often by a division-promoting factor, and they must be provided with adequate nutrients to sustain their own synthetic mechanisms. It is not surprising that there are variations in the ease with which division can be stimulated and in the raw materials appropriate to the necessary metabolic processes within the cells. What does emerge clearly is that the organized pattern of development that leads to the production of a whole plant cannot be ascribed to the external medium. Rather it has its origin in the intact genetic complement of the cells. The external milieu of the cell must trigger the expression of this pattern and, having triggered it, must sustain its development. It is doubtful that anything different can be said of the zygote and its milieu.

Plant regeneration and developmental regulation

When considering the conditions under which differentiated cells are able to express other potentialities, it must be emphasized that this ability is not confined to the artificial conditions of nutrient culture where the cells that produce whole plants are part of a callus mass or are freely suspended in culture. In fact, events that can be interpreted only as manifestations of the totipotency of differentiated cells are encountered with surprising frequency in plants as regeneration phenomena, and in some plants are of such importance that they constitute mechanisms of vegetative reproduction. This strongly suggests that the conditions that permit differentiated cells to express other potentialities do exist in many plants under certain circumstances, and it gives point to the enquiry as to why all cells do not reveal their latent capacities. Although this question cannot be answered on the basis of information presently available, some light may be shed upon it by an examination of the phenomena. From the extensive literature dealing with regeneration and vegetative reproduction in plants, only a few well-documented examples can be selected for discussion.

Patterns of regeneration

A recent investigation on the method of vegetative reproduction in the orchid *Malaxis paludosa* by Taylor (1967) has revealed an excellent example of the developmental potentialities that may be retained by differentiated cells. In this plant, terminal cells of the mature leaf regain meristematic activity and develop into tiny, egg-shaped embryos essentially like those produced in ovules, each surrounded by a sheathing jacket (Fig. 17.6). After detachment from the leaf, these foliar embryos develop into new plants, thus functioning essentially as does a normal seed embryo. This production of foliar embryos is a remarkable example of the retention of full zygotic potentialities by differentiated cells and of the ability of such cells to express these potentialities within the organization of the intact plant. There are, moreover, other well-known instances in which entire plantlets are initiated on leaves.

Many members of the Crassulaceae, a family of succulent plants, provide examples of the production of plantlets on leaves. In *Bryophyllum calycinum* Naylor (1932) has described the origin of complete plantlets

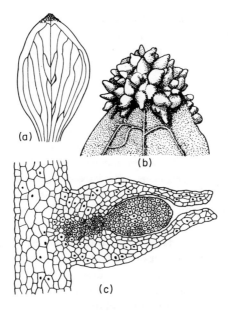

Fig. 17.6 Foliar embryos formed on the leaf of *Malaxis paludosa*. (a) A leaf of *Malaxis* showing the development of a cluster of foliar embryos at the extreme tip. (b) An enlarged view of the leaf tip with a well developed group of embryos each enclosed in a sheathing jacket of cells. (c) Longitudinal section of a partially developed embryo showing its relation to the surrounding jacket and to the tissues of the leaf tip. (a) ×5, (b) ×20, (c) ×120. (R. L. Taylor, 1967. Canadian J. Bot. **45**: 1553.)

from residual groups of meristematic cells that persist in the notches along the leaf margin. These meristematic groups first become evident when the leaf is a few millimeters in length, from tissue that constitutes the marginal meristem. While other cells around them expand and mature, those in the meristematic groups continue to divide and give rise to a pair of leaves, a shoot apex, and two roots. At this stage, in which the plantlet resembles but is not identical to an embryo, its development ceases and is not resumed until it is removed from the leaf, or until the leaf is detached. The development of such an embryoid from the partially differentiated cells of the leaf margin does not represent a reversion of fully mature cells as in the case of the orchid already discussed, but it does show that cells committed to the formation of a leaf lamina, and thus highly determined, retain the full zygotic potentialities and can express them in certain localized regions.

Other members of the same family show considerable variation in the extent of development of embryoids on leaves still attached to the plant. In *Kalanchoe daigremontiana* the plantlet produces several pairs of leaves and numerous roots while still attached to the parent plant. In other species, such as *Brynesia weinbergii*, the meristem remains dormant without the formation of any organ rudiments until the leaf is detached from the parent plant. Finally in *Crassula multicava* there is no evidence of persistent meristematic areas in the mature leaf; but following detachment of the leaf fully differentiated epidermal cells resume division and give rise to meristematic outgrowths that differentiate as plantlets. There are, in fact, many instances in which fully differentiated cells are reactivated and give rise to whole plants following injury or some other interruption of whole plant integrity. In *Saintpaulia ionantha*, for example, if a mature leaf is detached and placed on a moist substrate, epidermal cells near the base of the petiole are reactivated and organize shoot meristems. At the same time cells deeper in the tissue of the petiole give rise to root apices. Thus whole plants are produced although they do not arise from embryo-like structures. A similar type of plantlet initiation can occur from the lamina if a portion of this part of the leaf is explanted separately. Thus in some cases cells that in the normal course of events do not express their latent potentialities for further development will do so following injury or removal from the whole plant system.

There are also well-documented cases in which regeneration phenomena resulting from the reactivation of differentiated cells result in development of the more specialized meristem of a shoot or root, rather than producing an embryo or embryoid and thus a whole plant. An interesting example of this type of development is seen in the hypocotyl of *Linum usitatissimum* (flax). Link and Eggers (1946) have shown that these buds are initiated by divisions in individual epidermal cells or in groups of these cells. In the intact plant, development of these meristematic centers is arrested at an early stage, but occasionally they may develop as shoots. If, however, the plant is decapitated below the cotyledons, they regularly grow out and, in addition, many more primordia are formed.

Similarly, roots are often initiated from mature or partially mature tissues of the stem, or even the leaf. These may arise on intact plants, particularly at the nodes, but also in many cases without such localization. Their origin often follows injury and they are perhaps best recognized arising at the base of a severed stem in cuttings. In this way they provide the basis for the commonest method of horticultural propagation. In most cases these adventitious roots arise in the inner tissues of the stem close to the vascular system. Mature or partially mature parenchymatous cells undergo divisions to form a meristematic center which becomes organized as a root meristem and subsequently emerges at the surface of the parent stem. In some cuttings, roots originate from an unorganized callus that arises from the tissues of the stem near the cut end.

The regeneration of shoots or roots or both following injury or the severing of a particular organ is of importance not only from the point of view of horticultural propagation but from the standpoint of survival of the plant under natural conditions. Priestley and Swingle (1929), among many other workers, have investigated the development of shoots and roots that have propagative significance. One of the most interesting and least understood of these regeneration phenomena is the initiation of shoot buds on roots. Many plants initiate shoot buds on severed or damaged roots by the development of shoot meristems from differentiated cells of the internal tissues. Root fragments of such species can be used as propagules for intentional multiplication, and, under natural conditions, survival and even spread are enhanced. Some perennial weeds are extremely difficult to eradicate because their roots possess this property. Moreover, the production of root buds need not depend upon injury (Fig. 17.7); and there are many plants in which the formation of buds on undisturbed root systems provides a mechanism of vigorous spreading. An indication of the biological importance of this phenomenon comes from a recent survey of perennial dicotyledons in southern Saskatchewan. This revealed some 38 species of 19 different families in the prairie and forest border region that produce buds of shoots on their roots without any apparent injury or disturbance. This list included 11 of the most persistent and troublesome weeds of the region including leafy spurge (*Euphorbia esula*) which alone infests some 35,000 to 40,000 acres of otherwise valuable agricultural land in Canada. Another species, the aspen poplar (*Populus tremuloides*), reproduces on the prairies only by means of root buds, and large clumps covering many acres are believed to represent the development of original individual plants over periods of several thousand years.

It is thus evident that under a variety of conditions, differentiated or partly differentiated plant cells are capable of reverting to an embryonic condition and subsequently of expressing a new pattern of differentiation. The conclusion that any cell can express zygotic potentialities if its milieu is adequate to trigger and sustain this development is as valid for cells within a plant as it is for those growing in tissue cultures; the relevancy of the cell culture studies is thus strongly emphasized. However, it is clear that in the normal, intact plant very few cells actually are triggered to express alternate developmental patterns, and the development and maintenance of an organized plant body requires this limitation. The question of regulation, that is, of the integration of the plant body, is thus of prime importance. Although this regulation is far from being elucidated, there are some experiments that hint at the nature of the mechanism.

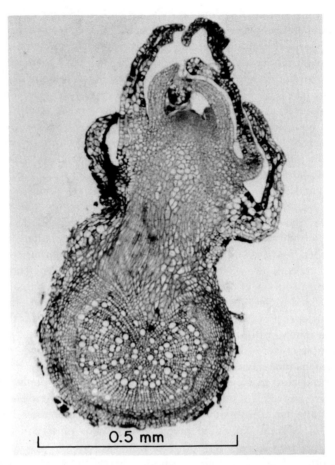

Fig. 17.7   Shoot bud regeneration from the root of *Viola adunca*. The root is shown in cross section in the lower part of the figure. In the upper part is the shoot with apical meristem, surrounding leaf primordia, scale leaves, and differentiating stem tissues. (M. V. S. Raju et al., 1966. Canadian J. Bot. **44** : 33.)

Biochemical mechanisms of developmental regulation

In horticultural practice it is well known that application of growth regulators of the auxin type often enhances the initiation of roots on cuttings and frequently promotes rooting in cuttings that will not form roots without it. Because of the practical significance of this phenomenon, it is not surprising that much research has been devoted to it. In the present context, however, what is significant is that it implicates a biochemical mechanism in the maintenance of integration in the plant, and suggests that when this is disturbed, regeneration may be the result.

Auxin also influences bud formation in a number of species but with rare exceptions this is an inhibitory effect. Thus in horseradish (*Armoracia rusticana*) Dore and Williams (1956) found that application of auxin to root slices promotes root initiation and inhibits bud formation, and these two responses increase with increasing concentrations of the hormone. Other substances are known that promote bud formation in various cases. Skoog and Tsui (1948), for example, have found a strong bud-promoting effect of adenine on stem segments of *Nicotiana tabacum* (tobacco) in sterile culture. An even more striking effect upon bud formation by the substance kinetin was shown in later experiments by Skoog and Miller (1957) on cultured tobacco stem segments and callus tissues. What emerged clearly from these experiments, however, was the importance of the relative concentrations of interacting substances rather than the absolute concentration of any one substance in regulating organ differentiation. By adjusting the relative concentrations of auxin and kinetin in the culture medium it was possible to induce bud formation, root formation or the growth of undifferentiated callus; and, within limits, the effect of a change in concentration of one substance could be countered by a corresponding change in the concentration of the other. However, in this and other experimental systems other substances have been found to play a role in controlling shoot and root formation in explanted tissues. In *Plumbago*, Nitsch and Nitsch (1967) have shown that in addition to an auxin and a cytokinin, adenine was necessary for organ formation on isolated stem segments and that gibberellins, several amino acids and amides, and the concentration of sugar profoundly influenced the type and the intensity of the developmental response.

The complexity of the developmental interaction between the tissues of the plant and the stimuli that act on them is particularly well demonstrated in a recent study by Aghion-Prat (1965) of regeneration in excised segments from the stem of flowering tobacco plants. Segments isolated from the lowest part of the stem and grown in a simple culture medium produced only undifferentiated callus. Those taken higher on the vegetative part of the stem formed vegetative buds in culture. Segments isolated from the base of the inflorescence formed shoots that either remained vegetative, or, after a short period of growth, developed a terminal inflorescence and flowers. Those explanted from the uppermost internodes of the inflorescence produced flowers directly with no intervening period of vegetative growth. These results seem to indicate a gradient of determination of cells of the plant axis toward reproductive development. However, even in explants from the terminal internodes of the inflorescence developmental expression could be modified, and when sugar was omitted from the medium, or present in low concentrations, only vegetative shoots were regenerated.

Another question to be kept in mind in considering organ formation is whether the triggering of development simultaneously determines the kind of organ that will be produced or whether the two processes are separate and perhaps independently controlled. An illustration of this problem is provided in a study by Bonnett and Torrey (1966) of the initiation of lateral roots and shoot buds on cultured root segments of *Convolvulus arvensis*. At the time of initiation the primordia of both types of organs were morphologically indistinguishable and only subsequently did they become recognizable as root or shoot apices as a result of differences in rate of development and the orientation of cell divisions. Both the number and kind of lateral organs produced on the root segments could be influenced by the constituents of the culture medium, but the implication of the morphological study was that the influences governing initiation of development in a regeneration site and those governing determination as shoot or root meristem were not exerted at the same time.

General comment

These experiments, and many others of a similar type, are beginning to give substance to the statement that any cell can express its latent potentialities if its milieu is such as to trigger and sustain its development. In the organized and integrated plant the suitable cell environment is apparently not often encountered, and therefore differentiated cells tend to remain in a stable mature condition and are mitotically inactive. The development of an organized plant body requires that this be so. However, under a variety of conditions, some of which occur spontaneously and some of which follow injury, the right combination of factors is achieved to trigger and sustain renewed development; and previously mature cells may then express their latent potentialities. In doing so, the full developmental potential of the cell may be expressed and an entire plant is produced. In other cases, however, a less complete expression is evoked and only the individual meristems of shoot or root are initiated. Why triggering in some cases should result in the whole developmental course being repeated and in other cases only a portion of this course is not clear and remains an exciting area for future investigation. Possibly the cellular environment during the initial stages of regeneration exerts a determining effect on the fate of the developing structure. If so, it is possible to visualize the meristems of the plant in normal development as representing the partially differentiated derivatives of the zygote, which in the embryo become determined as shoot- or root-forming structures, and are then maintained at this particular level of differentiation by the persistence of a specific set of factors.

The evidence at present available strongly suggests that the factors involved in regeneration are probably numerous and may vary in different cases. Attempts to interpret the phenomena of regeneration in terms of single specific substances or of simple interactions of a few substances have not been notably successful when applied to the whole organism and they do not seem likely to provide meaningful answers to the problems of plant morphogenesis. It seems much more probable that regulation, and consequently regeneration, are mediated through the total environment of the cell; and this conclusion is equally applicable to the zygote and to the differentiated cells in the mature plant body.

REFERENCES

Aghion-Prat, D. 1965. Néoformation de fleurs *in vitro* chez *Nicotiana tabacum* L. Physiol. végét. **3**:229–303.

Bonnett, H. T. Jr., and J. G. Torrey. 1966. Comparative anatomy of endogenous bud and lateral root formation in *Convolvulus arvensis* roots cultured *in vitro*. Am. J. Botany **53**:496–507.

Dore, J., and W. T. Williams. 1956. Studies in the regeneration of horseradish. II. Correlation phenomena. Ann. Botany [N. S.] **20**:231–249.

Haberlandt, G. 1902. Kulturversuche mit isolierten Pflanzenzellen. S. B. Akad. Wiss. Wien, Math-naturw. Kl. **111**:69–92.

Halperin, W. 1966. Alternative morphogenetic events in cell suspensions. Am. J. Botany **53**:443–453.

Halperin, W., and D. F. Wetherell. 1964. Adventive embryony in tissue cultures of the wild carrot, *Daucus carota*. Am. J. Botany **51**:274–283.

Jones, L. E., A. C. Hildebrandt, A. J. Riker, and J. H. Wu. 1960. Growth of somatic tobacco cells in microculture. Am. J. Botany **47**:468–475.

Konar, R. N., and K. Nataraja. 1965. Experimental studies in *Ranunculus sceleratus* L. Development of embryos from the stem epidermis. Phytomorphology **15**:132–137.

Link, G. K. K., and V. Eggers. 1946. Mode, site, and time of initiation of hypocotyledonary bud primordia in *Liunm usitatissimum* L. Botan. Gaz. **107**:441–454.

Muir, W. H., A. C. Hildebrandt, and A. J. Riker. 1958. The preparation, isolation and growth in culture of single cells from higher plants. Am. J. Botany **45**:589–597.

Naylor, E. 1932. The morphology of regeneration in *Bryophyllum calycinum*. Am. J. Botany **19**:32–40.

Nitsch, C., and J. P. Nitsch. 1967. The induction of flowering *in vitro* in stem segments of *Plumbago indica* L. I. The production of vegetative buds. Planta. **72**:355–370.

Priestley, J. H., and C. F. Swingle. 1929. Vegetative propagation from the standpoint of plant anatomy. U. S. Dept. Agric. Tech. Bull. **151**:1–98.

Reinert, J. 1967. Some aspects of embryogenesis in somatic cells of *Daucus carota*. Phytomorphology **17**:510–516.

Skoog, F., and C. O. Miller. 1957. Chemical regulation of growth and organ formation in plant tissues cultured *in vitro*. Symp. Soc. Exptl. Biol. **11**: 118–131.

Skoog, F., and C. Tsui. 1948. Chemical control of growth and bud formation in tobacco stem segments and callus cultured *in vitro*. Am. J. Botany **35**:782–787.

Steward, F. C., M. O. Mapes, A. E. Kent, and R. D. Holsten. 1964. Growth and development of cultured plant cells. Science **143**:20–27.

Steward, F. C., M.O. Mapes, and K. Mears. 1958. Growth and organized development of cultured cells. II. Organization in cultures grown from freely suspended cells. Am. J. Botany **45**:705–708.

Taylor, R. L. 1967. The foliar embryos of *Malaxis paludosa*. Can. J. Botany **45**: 1553–1556.

Vasil, V., and A. C. Hildebrandt. 1965a. Growth and tissue formation from single, isolated tobacco cells in microculture. Science **147**:1454–1455.

———— 1965b. Differentiation of tobacco plants from single isolated cells in microcultures. Science **150**:889–892.

Index

293